T0302209

Applied Linear Regression for Longitudinal Data

This book introduces best practices in longitudinal data analysis at an intermediate level, with a minimum number of formulas without sacrificing depths. It meets the need to understand statistical concepts of longitudinal data analysis by visualising important techniques instead of using abstract mathematical formulas. Different solutions such as multiple imputation are explained conceptually, and consequences of missing observations are clarified using visualisation techniques. Key features include the following:

- Provides datasets and examples online
- Gives state-of-the-art methods of dealing with missing observations in a nontechnical way with a special focus on sensitivity analysis
- Conceptualises the analysis of comparative (experimental and observational) studies

It is the ideal companion for researchers and students in epidemiological, health, and social and behavioural sciences working with longitudinal studies without a mathematical background.

Frans E.S. Tan is an associate professor (retired) of methodology and statistics at Maastricht University, The Netherlands.

Shahab Jolani is an assistant professor of methodology and statistics at Maastricht University, The Netherlands.

Chapman & Hall/CRC

Texts in Statistical Science Series
Joseph K. Blitzstein, *Harvard University, USA*
Julian J. Faraway, *University of Bath, UK*
Martin Tanner, *Northwestern University, USA*
Jim Zidek, *University of British Columbia, Canada*

Recently Published Titles

Probability, Statistics, and Data
A Fresh Approach Using R
Darrin Speegle and Brain Claire

Bayesian Modeling and Computation in Python
Osvaldo A. Martin, Raviv Kumar and Junpeng Lao

Bayes Rules!
An Introduction to Applied Bayesian Modeling
Alicia Johnson, Miles Ott and Mine Dogucu

Stochastic Processes with R
An Introduction
Olga Korosteleva

Introduction to Design and Analysis of Scientific Studies
Nathan Taback

Practical Time Series Analysis for Data Science
Wayne A. Woodward, Bivin Philip Sadler and Stephen Robertson

Statistical Theory
A Concise Introduction, Second Edition
Felix Abramovich and Ya'acov Ritov

Applied Linear Regression for Longitudinal Data
With an Emphasis on Missing Observations
Frans E.S. Tan and Shahab Jolani

For more information about this series, please visit: https://www.routledge.com/Chapman--HallCRC-Texts-in-Statistical-Science/book-series/CHTEXSTASCI

Applied Linear Regression for Longitudinal Data

With an Emphasis on Missing Observations

Frans E.S. Tan
Shahab Jolani

CRC Press
Taylor & Francis Group
Boca Raton London New York

CRC Press is an imprint of the
Taylor & Francis Group, an **informa** business

A CHAPMAN & HALL BOOK

First edition published 2022
by CRC Press
6000 Broken Sound Parkway NW, Suite 300, Boca Raton, FL 33487-2742

and by CRC Press
4 Park Square, Milton Park, Abingdon, Oxon, OX14 4RN

CRC Press is an imprint of Taylor & Francis Group, LLC

© 2023 Taylor & Francis Group, LLC

ISBN: 978-0-367-63431-5 (hbk)
ISBN: 978-0-367-63937-2 (pbk)
ISBN: 978-1-003-12138-1 (ebk)

DOI: 10.1201/9781003121381

Typeset in Palatino
by SPi Technologies India Pvt Ltd (Straive)

Access the Support Material: https://www.routledge.com/Applied-Linear-Regression-for-Longitudinal-Data-With-an-Emphasis-on-Missing/Tan-Jolani/p/book/9780367634315

To Jacqueline, Indji and Ailin

To Maryam, Ava and Tara

Contents

Preface

This book is a product of more than a decade experience in teaching multi-level analysis of longitudinal data (MALD) to PhD students and researchers in health, (bio)medical, social and behavioural sciences. Basically, the MALD course covers only linear regression methods, and extensions to nonlinear (e.g. logistic) models are not part of the course.

This book covers advanced methods for the analysis of longitudinal data, where the observations are correlated and/or subject to missingness. The standard regression method that assumes uncorrelated errors typically leads to biased estimates of regression parameters with incorrect standard errors and thus are inappropriate.

Special attention is given to the analysis of longitudinal intervention and life-event studies, where the objective is to evaluate a treatment or life-event effect. Several statistical methods to deal with missing observations are presented, depending on the type of missing data mechanism and whether the dependent variable (outcome), the independent variables (covariates) or both are partly missing.

Chapter 1 introduces the scientific framework of linear regression analysis and the underlying theory of missing data methods. Chapter 2 starts with a brief review of standard linear regression model, and the notation and terminology of multilevel linear models are introduced. In addition, this chapter reviews simple and advanced methods for handling missing observations. The material in Chapters 3-4 forms the heart of multilevel analysis. Various examples are used to introduce random-effects and marginal models and to explain steps of model building in longitudinal data. Suggestions are given on how to deal with missing data problems when considering imputation strategies. Chapter 5 compares the analysis of covariance (ANCOVA) and gain-score approach in pre/post measurement designs. To address the problem of missing observations, sensitivity analysis via multiple imputation is demonstrated. Chapters 6 and 7 serve as case-studies to perform a full analysis on longitudinal data in observational and experimental studies, respectively.

This book can be used as a text for a 12-week course. We recommend starting with a brief introduction to the statistical software/packages that will be used during the course. The book's companion website offers a manual to perform the analyses in SPSS and R. Although the theory is explained independently of any statistical package, SPSS and R were mainly used to produce tables and figures. We also provide data files in the SPSS system-file format that may be used (not necessarily in SPSS) when working on the assignments. An introduction in SPSS (or R) can be covered in 1 week.

Chapter 1 and 2 can be taught in 2 weeks including a discussion of the assignments. Furthermore, each of the Chapters 3-6 can also be taught in about 2 weeks, including a repetition of the preceding chapters and a discussion of the assignments. Finally, Chapter 7 can be covered in 1 week.

Basic knowledge of linear regression analysis and testing theory would be an advantage for a smooth understanding of the topics of this book. In fact, we have used it in courses where participants already had some experience analysing data with regression methods.

Acknowledgements

We are indebted to the Department of Methodology and Statistics at Maastricht University, The Netherlands, for facilitating an environment to write this book. We thank all colleagues and students who ever followed the MALD course and inspired us to even start this project. Their comments during and after the course were extremely useful.

Special thanks to Miranda Janssen for developing the companion website and a close collaboration in this project. We wish to thank David Grubbs and Lara Spieker, Statistics Editors at Chapman & Hall/CRC for their encouragement and patience during this project. Jessica Poile and Curtis Hill have been very helpful in facilitation of this project. Finally, several people read and commented on parts of the book. Many thanks to Martijn Berger for reviewing the manuscript and providing many helpful suggestions. We also thank Bjorn Winkens, Gerard van Breukelen, Math Candel, Jan Schepers, Sophie Vanbelle, Alberto Cassese, Francesco Innocenti, Andrea Gabrio, Gavin van der Nest and Monique Reusken for their insightful feedback. This has been very helpful to clarify the ideas.

Frans E.S. Tan & Shahab Jolani
Maastricht, 2022

Short Description of Research and Simulation Studies

Cross-sectional Nutrition study (first used in Section 2.1.1)

datafile: Nutrition.sav

As a part of a large study, a random sample of 104 school children was collected, divided in 56 rural and 48 urban children (Greenberg, 1953; Tan et al., 2008). There are no missing observations.

Relevant variables are:

- *School*: school district categorised as rural and urban,
- *Age*: child age in months,
- *Length*: child body length in cm.

Cross-sectional Nutrition study with missing observations (first used in Section 2.7.1)

datafile: Nutrition_missing.sav

A sample of children was randomly selected (about 20%) and their body length was set to missing. The other variables were fully observed.

Longitudinal Violent-behaviour study (first used in Section 2.3.1)

datafile: Alc_violent.sav

A simulated longitudinal dataset to study the relationship between violent behaviour and alcohol consumption. The data consist of five subjects each of which was measured five consecutive years. Because all subjects were not measured at the same year, the design is called 'unbalanced design'.

The dataset is generated such that there is a positive correlation between violent behaviour and alcohol consumption. To do so, subjects who drink more were also measured on a later occasion and also violent behaviour increases over time. Actually, the positive relationship between violent behaviour and alcohol consumption is due to a time effect: violent behaviour increases with time and people drink more over time.

Relevant variables are:

- *Subject*: subject identifier,
- *Time*: year of study (1950–1958),
- *Alcohol*: alcohol consumption in glasses per week,
- *Violent*: violent behaviour measured in a ten-points scale questionnaire.

Longitudinal Proximity study (first used in Section 2.4.1)

datafile: Teacher.sav, Teacher_wide.sav

In the Proximity study (Brekelmans, Wubbels, and Créton, 1990), a total of 50 teachers (36 males and 14 females) were evaluated on their interpersonal behaviour in the classroom. The degree of closeness between a teacher and his/her students was measured yearly for a period of four years, starting from the baseline and proceeding three occasions. The data contain missing observations.

Relevant variables are:

- *Teacher*: Teacher identifier,
- *Occasion*: measurement occasion,
- *Proximity*: proximity score,
- *Gender*: Teacher gender categorised as male and female.

Longitudinal Growth study (first used in Section 3.1.2)

datafile: Growthdata.sav

This is a study of orthodontic growth of 11 girls and 16 boys, which was first analysed by Potthoff and Roy (1964). In this study, the distance from the

centre of the pituitary gland to the pterygomaxillary fissure was recorded for each subject at ages 8, 10, 12 and 14 years.
Relevant variables are:

- *Subj*: child identifier,
- *Sex*: child sex categorised as boy and girl,
- *Age*: child age in years,
- *Distance*: distance between pituitary and maxillary fissure in mm.

Longitudinal Alzheimer study (first used in Section 3.5.1)

datafile: Alzheimer.sav

The Alzheimer study was a clinical trial involving 344 patients with Alzheimer disease aiming at comparing placebo with two active treatments (Verbeke and Molenberghs, 2006, par.17.18). The outcome was dementia score measured at baseline (twice) and over six occasions after treatment. Baseline variables involved sex and age of patients. In this study, the dementia score had missing observations, but all baseline variables were fully observed.
Relevant variables are:

- *Patid*: patient identifier,
- *Treatment*: treatment group (0 = placebo, 1 = active treatment 1, 2 = active treatment 2),
- *Age*: patient age at baseline in years,
- *Gender*: patient gender categorised as male and female,
- *Week*: measurement time in weeks,
- *Alzheimer*: dementia score.

Longitudinal Salsolinol study (first used in Section 4.2.3)

datafile: Salsolinol.sav

In the Salsolinol study, two groups of subjects, one with moderate dependence on alcohol (six subjects) and the other with severe dependence on alcohol (eight subjects), had their salsolinol secretion levels measured on four consecutive days (Landau and Everitt, 2004). Primary interest is whether the groups evolved differently over time. In this study, there were no missing observations, and the design was balanced in time with equidistant time points. All salsolinol level measurements were log transformed due to skewness of the raw measurements.

Relevant variables are:

- *Id*: subject identifier,
- *Group*: = dependency group (0 = moderate dependence on alcohol, 1 = severe dependence on alcohol),
- *Time*: measurement day,
- *Salsolinol*: salsolinol secretion level in mmol,
- *Tsals*: log-transformed salsolinol level.

Longitudinal Beating the Blues study (first used in Section 5.2.1)

datafiles: LongBTB.sav, Ancova BTB.sav and BTB_wide.sav (wide data format)

Beating the Blues (BtB, Proudfoot et al., 2003) was a clinical trial designed to assess the effectiveness of an interactive programme using multi-media techniques for the delivery of cognitive behavioural therapy to depressed patients. In this study, patients with depression recruited in primary care were randomised to either the BtB program, or the Treatment as Usual (TAU). The outcome measure used in the trial was the Beck Depression Inventory II with higher values indicating more depression. The baseline measurements were measured before assigning the patients to either new or standard drug. The dependent variable measurements were made on five occasions:

- prior to treatment, and
- follow up at 2, 3, 5 and 8 months after treatment

In addition, the dataset included an additional variable that indicates whether the patient had an illness history of longer than 6 months.

Relevant variables are:

- *Subject*: subject identifier,
- *Duration*: duration of illness,
- *Treat*: treatment group (1 = BtB, 0 = TAU),
- *Month*: measurement month,
- *Depress*: depression score.

Longitudinal Well-being study (first used in Section 5.2.2)

datafile: Lifesubset.sav and Lifeancova.sav

Nieboer et al. (1998) studied the effect of loss and illness of the partner on the state of well-being in 263 respondents. In this prospective research study, there were 157 caregivers whose partner had an illness-incidence and 112 widow(er)s whose partner died. Three repeated measurements were taken on well-being (the higher the score, the better the well-being): a baseline measurement, a post measurement after 3 months and a follow-up measurement after 12 months. The baseline measurements were obtained before the life event occurred including well-being at baseline, gender and age of participants. These variables were considered as possible confounders.

Relevant variables are:

- *Subj*: subject identifier,
- *Gender*: respondent gender categorised as male and female,
- *Age*: respondent's age in years,
- *Group*: group categorised as caregiver and widow(er),
- *Time*: measurement time,
- *Well-being*: well-being score.

1

Scientific Framework of Data Analysis

1.1 Validation Modelling: Comparative Studies

An important aspect in epidemiological, social and behavioural, and health research is to compare two or more groups with respect to some outcomes. For example, in a study in which the effect of smoking on physical condition is evaluated, smokers and non-smokers may be compared with respect to a change in pulse rate (*CPR*) determined from measurements before and after an exercise. A problem arises if the two groups (smokers vs. non-smokers) differ from each other with respect to several group characteristics besides smoking status. As a result, the observed difference in *CPR* could be attributed to differences in weight distribution, for instance. Smokers might be thinner than non-smokers and a difference in average *CPR* could then be partly due to the difference in average weight. If the interest is to determine the net effect of smoking status on *CPR*, the researcher somehow needs to consider this weight difference. Besides the difference in weight, there might also be distributional differences with respect to other factors such as *Age* and *Gender*, which may be associated with smoking status and *CPR*. Such factors are also known as confounders (see, e.g., Tan, 2013). Matching methods is one way to take these factors into account (see Cochran, 1983, Rosenbaum, 2002, Rubin, 2006). Another way is to 'correct' for the observed difference in average *Weight* using advanced statistical methods, for example, the linear regression method.

Figure 1.1 shows a plot of *CPR* (vertical axes) against *Weight* (horizontal axis). As can be seen, the raw (unadjusted) difference in average *CPR* between smokers and non-smokers is equal to $\overline{CPR}_{smokers} - \overline{CPR}_{non-smokers}$. It should be noted that a standard *t*-test could be performed to compare both the groups. However, the average weight also differs between smokers and non-smokers (smokers are thinner than non-smokers on average). Moreover, *Weight* is (negatively) related to *CPR*. If there is no causal effect of smoking on body weight (see, e.g., Tan, 2013), then *Weight* is a confounder. Hence, the correct way to determine the net effect of smoking on *CPR* is to set *Weight* at an arbitrary fixed, but equal value for both smokers and non-smokers.

DOI: 10.1201/9781003121381-1

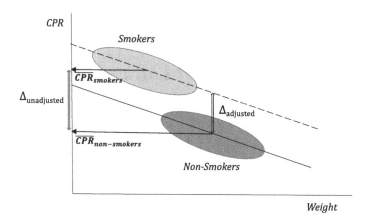

FIGURE 1.1
An illustration of the adjustment for a confounder in linear regression analysis.

Figure 1.1 depicts the difference in average *CPR* between smokers and non-smokers at a fixed but arbitrary value of *Weight*. This is called the 'adjusted' difference.

Note that the adjusted difference in *CPR* is smaller than the raw difference, that is the observed difference in *CPR* is partly due to a weight difference between smokers and non-smokers.

An adjusted difference could also be larger, depending on the relation between the dependent variable and the confounder and/or the average value of the confounder for each treatment group. For instance, in the example of smoking, the average weight among smokers is smaller than that of non-smokers. If, on average, smokers were heavier than non-smokers, the adjusted difference in *CPR* would be larger than the raw difference. Linear regression models used to account for these differences are elaborated in detail in Chapter 2.

A more complex situation is when one wishes to compare the dependent variable profiles between groups over time. In such cases, confounders may also change over time. Standard linear regression methods that will be described in Chapter 2 are then inappropriate, because these methods invalidate the estimate and/or standard error of the effect. More suitable methods are advanced regression models with correlated errors that accommodate the dependency of measurements over time. In Chapter 3 and the chapters that follow, we elaborate the linear regression models for correlated data in detail.

A situation that is often used in clinical research is when patients are randomly assigned to two or more treatment groups according to a randomisation procedure (see, e.g., Meinert, 2009). For example, consider a study about systolic blood pressure (*SBP*), wherein a standard drug is compared with a new drug. A research team designs a randomised clinical trial (RCT) and randomly assigns patients to one of the two treatment groups. Because of randomisation, the two groups are expected to be comparable regarding the distribution of all characteristics such as *Gender, Age* and *Education*. Hence, any difference in *SBP* post treatment is expected not to depend on these other factors, but only on the difference during the course of treatment. As we will observe in Chapters 5–7, this difference has consequences for modelling and analysing longitudinal data.

1.2 Predictive Modelling: Predicting a Future Outcome

Searching for important determinants of an outcome (dependent variable) is sometimes the main objective of a study. This is because the results may form the basis for future intervention studies or a statistical model with high prediction quality is desired for making reliable predictions for future patients. As an example, Daamen et al. (2017) carried out a cross-sectional study in five long-term care organisations in the southern parts of the Netherlands. About 4500 nursing home residents were involved. In order to develop a highly accurate prediction model, they considered 14 potential prognostic characteristics to predict the presence of heart failure among nursing home residents. Of these, seven predictors (or independent variables) were selected that showed an important contribution in predicting heart failure. In contrast to a validation analysis, wherein the main interest, for example, is in the treatment effect, all the seven independent variables are important. One might also be interested in the independent contribution of each of these seven prognostic factors in predicting the presence/absence of heart failure. Although the statistical models presented in this book can be used for prediction purposes, the applications in the book are mainly about validation modelling. All independent variables (except the comparison variable of interest) should be treated as nuisance variables that need to be considered as confounders or effect modifiers (Tan, 2013).

Interested readers are referred to, for example, studies by Kleinbaum et al. (2008), Steyerberg (2009) and James et al. (2017) for an introduction to predictive modelling.

1.3 Missing Observations

Missing observations are essentially inevitable, particularly in longitudinal studies, as individuals typically dropout for various reasons.

→ *We define a missing observation as a planned measurement that could have been observed, but it is missing.*

In a clinical study of hypertension, as an example, the score of blood pressure for a patient is missing (i.e. not measured) if the patient does not show up for a scheduled visit. In this chapter, we explicitly exclude the possibility of missingness due to death, as methods developed to handle missing observations in longitudinal data due to death are still open to debate (Biering et al., 2015).

The analysis of data in the presence of missing observations becomes particularly complicated simply because standard statistical methods are not designed to accommodate missing observations. An even greater problem is that missing observations may invalidate the results. In a clinical trial of aggression, as another example, the beneficial effect of an intervention could easily be overestimated if more aggressive children leave the trail prematurely.

In general, the methods to handle missing data can be categorised into the following four categories:

1. Deletion methods: these methods are simple solutions such as complete-case analysis or available-case analysis, which discard subjects with missing observations in different ways.
2. Weighting methods: these methods only use subjects with complete observations (i.e. those for which all data are observed), but weights for responding units are included in the analysis in an attempt to account for the non-respondents.
3. Imputation-based methods: missing values are filled in these methods, and the resultant completed dataset is then analysed as if there were no missing observations. An advanced imputation technique is multiple imputation, which reflects the added uncertainty due to the fact that the imputed values are not real values.
4. Model-based procedures: these methods build the analysis on the observed data and maximise the corresponding likelihood directly. Inference can be according to likelihood-based or Bayesian methods.

In this book, we largely focus on the last two categories, particularly on the state-of-the-art approaches, namely, multiple imputation and direct likelihood. For the purpose of comparison, single imputation and deletion methods will also be considered if necessary.

1.3.1 Missing Data Mechanisms

The process under which missing data are obtained (i.e. why certain values are observed while others are not) plays a crucial role in the validity of the missing data methods. Therefore, the missingness process, the so-called missing data mechanism, should be carefully investigated before applying any missing data methods to the data at hand. Following the study by Rubin (1976), modern statistical analysis of missing data distinguishes three types of missing data mechanisms, namely, *missing completely at random (MCAR)*, *missing at random (MAR)* and *missing not at random (MNAR)*. In the following section, we discuss each mechanism separately.

Missing completely at random: The missing data mechanism is MCAR if the chance of observing a score (e.g. *SBP*) is the same for all subjects. This means observing *the SBP* is totally unrelated to any data value, and hence, missing observations are due to pure chance. Obtaining a random sample from the grades of students in a statistical course is an example of MCAR because each student basically has the same chance of selection, and thus, the grades are observed (or not observed) by chance.

→ *Formally, MCAR implies that the probability that a unit provides data on a particular variable is not a function of any observed or unobserved data.*

In order to understand the MCAR concept better, let us define providing data on a particular variable as an event. Then, the MCAR mechanism implies that the occurrence of the event (i.e. whether the value is observed on a particular variable) does not depend on any observed or unobserved values. This definition clarifies that the MCAR mechanism is often too restrictive and hardly defensive for the data at hand.

If the reason for missing observations depends on some additional factors that are totally unrelated to the study at hand, we also refer to the missing data mechanism as MCAR (see Figure 1.2a). For example, the mechanism

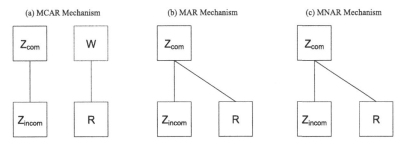

The boxes labelled Z_{com} and Z_{incom} represent fully and partially observed collections of variables, respectively. The box labelled R represents the missing data indicator, which takes 1 if the corresponding value in Z is observed and zero otherwise. The dashed box labelled W represents a collection of variables related to missingness but unrelated to variables in the study at hand.

FIGURE 1.2
A graphical representation of missing data mechanisms.

is MCAR if a patient does not show up for a follow-up visit because of mis-communication or because he/she has forgotten to attend. Although the probability that the follow-up visit is missed is a function of some additional variables/factors, these are unrelated to the study at hand, and therefore, we classify the missing data mechanism as MCAR. In sum, if it is known that additional factors cause missing data, but unrelated to the analysis model (particularly the dependent variable), then, the missing data mechanism can be considered as MCAR.

→ *When the missing data mechanism is* <u>MCAR</u>, *complete cases (i.e., units for which all data are observed) can be considered to be a random sample from the original population and therefore any analysis restricted to complete cases is valid.*

Nevertheless, the MCAR assumption can easily be violated in practice.

Missing at random: The MAR mechanism refers to situations in which the missingness process is completely explained by the observed data (see Figure 1.2b).

→ *Stated otherwise, if the probability that a unit provides data on a variable is only a function of the observed values, the missing data mechanism is MAR. This implies that conditional on the observed data, the missing data mechanism is not related to any missing value.*

For instance, in a study investigating blood pressure among students, if the blood pressure is more likely to be measured for boys than for girls, the missing data mechanism for blood pressure (conditional on *Sex*) can be classi-fied as MAR as long as *Sex* is fully observed and taken into account in the analysis. Note that *Sex* is related to missingness (and possibly related to the dependent variable) in this example.

Under the MAR mechanism, complete cases cannot be considered as a random sample from the original population, as opposed to the MCAR mechanism, because the distribution of complete cases corresponding to the analysis model is not the same as the analogue distribution for full data (i.e. when there were no missing observations). Hence, an analysis based on complete cases (i.e. complete-case analysis) is generally invalid under the MAR mechanism (Little and Rubin, 2020, p. 48). An exception is in cross-sectional studies, wherein the dependent variable has missing observations only (and independent variables are fully observed). We get back to this point in Chapter 2. In contrast to the cross-sectional studies, complete-case analysis is invalid in longitudinal studies even if only the dependent vari-able has missing observations. Examples of the MAR mechanism will be discussed extensively in subsequent chapters for cross-sectional and longitudinal designs.

An important implication of MAR mechanism is that the conditional distribution of the incomplete variable (e.g. blood pressure in our example), given the variable(s) causing missingness (e.g. *Sex*), is the same among observed and missing values. This means that

1. The distribution of missing blood pressure for boys is the same as the distribution of observed blood pressure for boys.

2. The distribution of missing blood pressure for girls is the same as the distribution of observed blood pressure for girls.

Equivalently, it can be said that the missing data mechanism for blood pressure is MCAR within categories of *Sex*; that is, for boys and girls separately, whether the blood pressure is (un)observed does not depend on observed or unobserved data.

The MAR mechanism is very appealing, particularly for imputation purposes, because the missing values can be imputed within each category (e.g. within categories of *Sex*), and then, the completed data set will be used for further analysis. Nevertheless, the MAR mechanism is an untestable assumption because the missing data that are needed to check this assumption are not available from the data at hand. For an elaboration, see Carpenter and Kenward (2013, pp. 14 and 15).

When the MAR mechanism is a plausible assumption, the process that creates missing data can be ignored when making inferences about the parameters of interest (e.g. regression coefficients). This is because the likelihood corresponding to the missingness process does not contain any information about the parameters of interest (see, for instance, Little and Rubin, 2020, p. 113). Hence, in longitudinal studies, for example, any likelihood or Bayesian analysis that ignores the missingness process is valid as long as the MAR mechanism is not violated. This concept is typically referred to as 'ignorable' in a missing data context. The precise definition of ignorability following the study by Rubin (1976) is when (1) the parameters governing the missingness process are distinct from the parameters in the analysis model (or they are *a priori* independent in Bayesian context), and (2) the missing data mechanism is MAR. Because the MCAR mechanism is a special case of MAR, any analysis ignoring the missing data mechanism is valid for MCAR as well. For this reason, these two mechanisms are often referred to as ignorable mechanisms. It should be noted that if the missing data mechanism is ignorable, it does not imply that cases with missing values can be removed from the analysis.

Missing not at random: If the probability that a unit provides data is a function of the values that are missing, even after conditioning on the observed values, the missing data mechanism is referred to as MNAR.

→ *Stated otherwise, if providing data on a variable also depends on unob-*
served and not only on observed values, then the missing data is MNAR.

In a clinical trial of depression, for instance, if the depression score at the end
of the trial is more likely to be missing for patients with a high depression
score (i.e. those who do not find the treatment effective), then the missing
data mechanism is MNAR. This is because the process that creates missing
values on depression depends on the score of depression itself (see Figure
1.2c), which is obviously not observed.

Under the MNAR mechanism, the missingness process must be considered
when inferences about the parameters of interest are made. Any likelihood
or Bayesian analysis that ignores the missing data mechanism is invalid and
produces biased estimates. For this reason, the MNAR mechanism is often
referred to as 'non-ignorable.' In general, any valid inference for MNAR data
requires specification of a model for the missingness process.

→ *However, the assumed model for the missing data mechanism is totally*
unverifiable from the data at hand because any evidence concerning the
relationship of missingness to the missing values is absent by definition.
Therefore, it is important to investigate the sensitivity of estimates to the
assumed models for the missing data mechanism.

Many authors have emphasised the importance of conducting sensitivity analysis (see, e.g., Little and Rubin, 2020, chapter 15; Molenberghs and
Kenward, 2007).

1.3.2 Patterns of Missing Data

Depending on the location of missing values, different patterns of missing
data can occur. Let us define the whole data as a rectangular matrix such that
each row represents an individual or a subject, and each column contains
measurements of a particular variable (e.g. the first column shows the height
of subjects; the second column includes *the SBP* measurement of subjects,
and so on). We distinguish three general types of missing data patterns as
follows.

The simplest pattern of missing data is the univariate pattern, which
implies that only one variable is not fully observed (i.e. one column in the
data matrix), while the other variables are fully observed for all individuals (see, Figure 1.3, univariate missingness). This type of pattern may occur
more frequently in cross-sectional studies than in longitudinal studies.
For instance, in an observational study, a researcher investigates the effect
of smoking on forced expiratory volume (*FEV*) adjusted for *Age* among

chronic obstructive pulmonary disease (COPD) patients. If smoking status and *Age* are fully observed for all patients, but *FEV* is only measured for some patients, the pattern of missing data is univariate. The analysis of data with a univariate pattern is simple and does not require any iterative optimisation algorithms in order to deal with missing values. This univariate pattern of missing data is particularly appealing for the purpose of imputation because the imputation algorithm will converge immediately after the first iteration. Basically, the information available in the complete cases is sufficient to obtain imputations for the missing values without iteration.

A slightly more complicated pattern of missing data is the monotone pattern, which visually resembles a staircase (see Figure 1.3, monotone pattern). This pattern is very common in longitudinal studies, wherein subjects drop out of the study and then never return. Suppose that in a randomised study comparing two treatments, namely, eye movement desensitisation and reprocessing (EMDR) and cognitive behavioural therapy (CBT), patients with post-traumatic stress disorder (PTSD) have been offered four scheduled sessions. If our patient John misses the second session, then the monotone pattern implies that he will miss the subsequent sessions too (i.e. 3 and 4). Therefore, John will have missing measurements in sessions 2–4.

Analysis of missing data with a monotone pattern is straightforward because the estimation process can be decomposed into separate parts, and therefore, the complexity of the estimation procedure is greatly reduced, even without a need for iterative estimation algorithms (see, Little and Rubin, 2020, pp. 151 and 152).

The last pattern that we discuss here is the non-monotone pattern or intermittent pattern. Figure 1.3 (right panel) visualises a non-monotone pattern

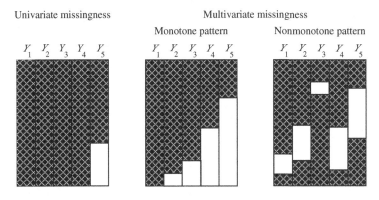

FIGURE 1.3
Different missing data patterns. Blank is missing.

of missing data for a data set with five variables. This type of pattern can occur in both cross-sectional and longitudinal studies and is the most common type of missing data pattern in practice. In the clinical trial of the PTSD patients, for instance, the pattern of missing data is non-monotone if John returns and participates in sessions 3 and 4 despite skipping session 2. In the FEV example among COPD patients, the missing data pattern can also be non-monotone if the explanatory variables smoking status and *Age* have missing values in addition to the dependent variable *FEV*.

The non-monotone pattern of missing data complicates the statistical analysis of data at hand because the estimation process cannot be factorised into separate parts anymore, and hence, the estimation procedure is complex. Usually, iterative estimation algorithms are needed to obtain the maximum likelihood estimates of the parameters of interest (e.g. regression coefficients), which is a difficult task and is not readily accessible by standard statistical packages. For these reasons, imputation-based approaches became very appealing for the non-monotone pattern of missing data, as the resultant completed data are typically analysed as if there were no missing values.

1.3.3 Randomised versus Nonrandomised Studies

Missing data can raise different issues in randomised studies than in nonrandomised studies. In randomised trials, the main goal is to find the causal effect of the treatment on the outcome. For this reason, the treatment allocation is purely random, which results in balancing the subjects with respect to their background characteristics such as *Age* and *Sex*. An important consequence of such a practice is the independence of the treatment variable from all other baseline variables (which are usually measured before randomisation) in the analysis model. Therefore, the performance of the developed methods for handling missing data is not necessarily equivalent in experimental and non-experimental studies. As an example, Groenwold et al. (2012) showed that, for missing observations in an independent variable, inclusion of the missing indicators in the analysis model as an additional independent variable produces an unbiased estimate of the treatment effect in randomised trials, while it produces a biased estimate in observational studies. In subsequent chapters, we will distinguish and discuss methods suitable for dealing with missing data for both randomised and nonrandomised studies.

1.3.4 Concluding Remarks

Missing observations are the rule rather than the exception in many different empirical studies and pose great analytical challenges for data analysis. Inappropriate handling of missing observations may invalidate any inference from data by, for example, overestimation of the benefits or underestimation of the harms of an intervention on patients.

After the seminal paper on missing data by Rubin (1976), the importance of missing data mechanisms has been recognised, which in turn resulted in the development of advanced methodologies for handling missing data. In this book, we put more emphasis on modern methods such as direct maximum likelihood and multiple imputation. In addition, it is important to distinguish between models that assume the missing data mechanism is MAR and models that rely on the MNAR assumption. The former ignores the missingness process during the estimation procedure, while the latter includes the missingness mechanism as an essential part in the estimation procedure. Both models for MAR and MNAR data, however, rest generally on (inherently) unverifiable assumptions, and hence, sensitivity of the results to violations from these assumptions must be carefully examined in practice. As a result, more recent statistical literature on missing data is directed towards the importance of sensitivity analysis (see, Little and Rubin, 2020, chapter 15 and references therein).

1.4 Assignments

1.4.1 Assignment

Figure 1.1 shows an example for which the observed *CPR* is partly due to weight difference between smokers and non-smokers.

- a. Argue the assumption that there is no causal effect of smoking on body weight.
- b. Consider another example from real life for which the raw difference is smaller than the adjusted difference.
- c. What can be said about the relation between the dependent variable of your example and *the confounder* on the one hand and between *the dependent variable* and *the risk factor* on the other hand? Show this by means of a plot such as Figure 1.1.

1.4.2 Assignment

Discuss an example (probably your own research study) for which a validation model should be used rather than a prediction model or *vice versa*.

1.4.3 Assignment

Describe the differences between MCAR and non-MCAR (i.e. MAR and MNAR) mechanisms. Is an MCAR mechanism more plausible than a non-MCAR in longitudinal studies?

1.4.4 Assignment

Considering a real-life application you are familiar with, describe the pattern of missing data and discuss whether the missing data mechanism can be classified to MCAR, MAR or MNAR.

2

Revisiting and Shortcomings of Standard Linear Regression Models

2.1 Standard Linear Regression Modelling

2.1.1 Formulation of the Research Question: The Cross-Sectional Nutrition Study

As a part of a large study, a random sample of 104 schoolchildren was collected, which was divided into 56 rural schoolchildren and 48 urban schoolchildren (Greenberg, 1953; Tan et al., 2008). Three variables are involved in this study, which are described as follows:

- *School* with categories rural (coded 0) and urban (coded 1),
- *Age* measured in months, and
- Body *Length* measured in cm.

Typical scientific questions are

1. How are *Age* and *Length* related (associated)?
2. What is the difference between rural and urban children with respect to average *Length* and does it depend on *Age*?

Regarding the first question, one might be thinking of the Pearson correlation, which is explained conceptually in the next section, but it is important to realise to which relationship we are exactly referring. The second question is probably more relevant for the research study, that is we might be interested in making a comparison between the two groups of schools with respect to the body length of children. It is not, however, the intention to compare all children one by one, but merely on average. One could, for example, question whether the average body length of the rural children differs from the average body length of the urban children and whether this difference depends on *Age*.

DOI: 10.1201/9781003121381-2

TABLE 2.1

Nutrition Study: Descriptive Statistics for Urban/Rural Children

		N	Minimum	Maximum	Mean	Std. Deviation
Age		104	109.00	142.00	129.55	8.90
Length		104	131.00	165.30	142.69	7.05

School		N	Minimum	Maximum	Mean	Std. Deviation
Rural	*Age*	56	121.00	140.00	133.07	6.36
	Length	56	131.00	149.50	141.67	5.95
Urban	*Age*	48	109.00	142.00	125.44	9.70
	Length	48	132.70	165.30	143.87	8.04

Table 2.1 represents some descriptive statistics of the two school types. The average *Age* is 129.55 months (ca. 11 years) and the average body length is 142.69 cm. When comparing rural with urban children, we also notice that the rural children are somewhat older (7.63 months older on average), but still slightly shorter (about 2 cm on average). This could explain why the researchers investigated this phenomenon as a proxy of different nutritional habits (Greenberg, 1953). The analysis is typically started by constructing a scatter plot and calculating the Pearson's product-moment correlation (which we denote as correlation hereafter). However, interpretation of the scatter plot requires a good grasp of the concept of correlation.

2.1.2 The Concept of Pearson (Product-Moment) Correlation

The correlation between two variables X and Y is defined as the average product of standardised scores.

$$r_{X,Y} = \frac{1}{n-1} \sum_{i=1}^{n} Z_{X_i} \times Z_{Y_i}, \tag{2.1}$$

where Z_{X_i} (and Z_{Y_i}) is the standardised score of X_i (and Y_i) for child or subject i with average zero and standard deviation 1. Note that the sum is divided by the degree of freedom (equal to n-1) instead of the number of observations. However, for the ease of presentation, we consider this as average. The standardised score Z_{X_i} is dimensionless and scale free and is defined as

$$Z_{X_i} = \frac{\left(X_i - \bar{X}\right)}{SD(X)},$$

$$\bar{X} = \frac{1}{n} \sum_{i=1}^{n} X_i, \; var(X) = \frac{1}{n-1} \sum_{i=1}^{n} \left(X_i - \bar{X}\right)^2, SD(X) = \sqrt{var(X)}. \tag{2.2}$$

→ *Standardised scores Z_{X_i} $i = 1, \ldots, n$ are centred around the average value and normalised. Consequently, Z_{X_i} has average 0 and standard deviation 1.*

The correlation between X and Y can be conceptualised by Figure 2.1, which is a scatter plot of data points with origin $(\overline{X}, \overline{Y})$. Let us divide the area of the coordinates into four quadrants. The first quadrant is the upper right corner, the second quadrant is the upper left corner, the third quadrant is the lower left corner and the fourth quadrant is the lower right corner. Note that Z_{X_i} expresses the deviation of the score of the ith subject from the average value $\overline{X}$ in units of the corresponding standard deviation SD(X).

The contribution of each observation pair (X_i, Y_i) to the correlation of Equation 2.1 is $Z_{X_i} \times Z_{Y_i}$. On the one hand, in the first quadrant, it is positive because, in the X-direction and in the Y-direction, the deviation of the individual scores from the average is positive. Consequently, the standardised scores Z_{X_i} and Z_{Y_i} of all observations in the first quadrant are positive. Along the same reasoning, the contribution of all observations in the third quadrant is also positive, while the contribution of all observations in the second and fourth quadrant is negative. If there are more points with large deviations from the origin in the first and third quadrant than in the second and fourth quadrant, the correlation will be positive. On the other hand, if the scatter

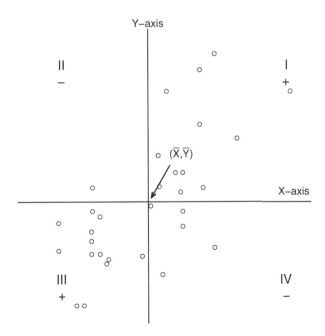

FIGURE 2.1
Visual interpretation of correlation coefficient.

plot stretches from the upper left corner to the lower right corner, then the correlation will be negative.

For the Pearson product-moment correlation, the following properties are valid:

1. It measures how close the observations are clustered around a straight line.
2. It is bounded between −1 and 1, i.e., $-1 \leq r_{X,Y} \leq 1$.
3. If the positive differences of X from its average are associated with positive differences of Y from its average, then $r_{X,Y}$ is positive.
4. $r_{X,Y}$ is dimensionless and does not depend on the scale of the measurement.

Sometimes, the covariance instead of the correlation between X and Y is calculated. The covariance is just an unstandardised version of the correlation. However, for the purpose of interpretation, working with correlations is more advantageous. The covariance between X and Y is defined as

$$COV(X,Y) = \frac{1}{n-1}\sum_{i=1}^{n}(X_i - \bar{X}) \times (Y_i - \bar{Y}) = r_{X,Y} \times SD(X) \times SD(Y). \quad (2.3)$$

For the nutrition study, Figure 2.2 shows the scatter plot of *Length* against *Age*. We expect a positive correlation because older children typically tend to be taller. This is confirmed by the scatter plot and the correlation that is equal to $r_{X,Y} = 0.31$. The relationship is, however, not perfect. In particular, for each *Age* value, there is variability in *Length*. The reverse is also true. For each *Length* value, there is variability in *Age*. There is also symmetry here, that is the correlation between X and Y is the same as the correlation between Y and X.

The conclusion that older children tend to be taller is not really a surprise! It is therefore more informative to know how exactly *Length* depends on *Age*, that is if we know the age of a child, what can we say about its body length in general? Of course, it does not make sense to do this for every child because of the existing variability in body length. There may be several factors other than *Age*, which can explain why one child is taller than the other, for example, differences in genetic background and variations in nutrition or environment. It would be interesting to know how the average body length of all children in our target population depends on the age of the children or groups (urban/rural) of children. In scientific terms, we would like to predict the average body length from information about the age of the children or equivalently to determine the effect of *Age* on *Length* (possibly for each group (urban/rural)).

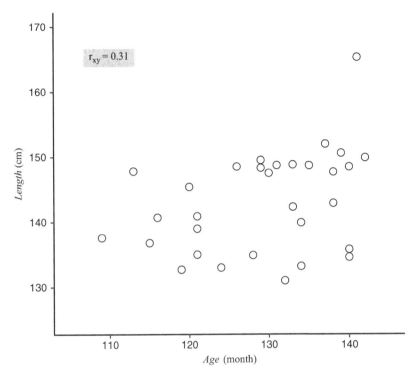

FIGURE 2.2
Scatterplot of *Length* against *Age*.

2.1.3 Specification of a Standard Linear Regression Model for the Nutrition Study

A well-known method to determine the effect of *Age* on *Length* is to use a linear regression model of Y on X.

→ *Note, that a linear regression analysis covers the determination of the (approximate) average Y value for a given X value (or conditional on X).*

Although we are analysing the relationship between *Age* and *Length* using a linear regression model, there is a conceptual difference between correlation and regression.

→ *Correlation is a symmetric concept, i.e., if X is correlated to Y, then Y is also correlated to X by the same amount. Regression, on the other hand, is an asymmetric concept. The regression of Y on X is generally not equal to the regression of X on Y.*

Generally, the average Y-value for a given X is not equal to the average X-value for a given Y, and the corresponding regression lines Y on X and X on Y will also not be equal. In our case, we are interested in the regression of Y on X. The variable Y is called the dependent variable because its value depends on variables such as X, which we denote as an independent variable. Other synonyms for the independent variable are covariate, predictor, classifier, (risk) factor or determinant.

The general idea behind a linear regression analysis is to summarise the scatter plot by means of a straight line that can be interpreted as (approximate) average values of *Length* (Y) for different values of *Age* (X) in our example. In practice, a linear regression analysis can be performed if one is interested in the average body length of children given that the *Age* is known.

Moreover, a linear regression analysis is worthwhile if one is interested in how the average body length changes when a child gets older. Unfortunately, we have no repeated measurements of the length of the children at different ages so that we cannot make inferences about 'a change in body length when the child gets older.' To do this, we will need a longitudinal design with repeated measurements of body lengths for each child getting older. In the present case, however, we have the so-called cross-sectional design, in which each child is measured only once. Thus, there are no repeated measurements. Chapter 3 and the chapters that follow deal with the analysis of repeated/longitudinal data.

For the cross-sectional nutrition study, one could be interested in the question 'how will *Length* differ on average if two groups of children are compared who differ one month from each other?' To answer this, the so-called regression line must be constructed.

If we want to summarise the scatter plot by determining the average *Length* for each *Age* category, then we can imagine dividing the horizontal axis (X-axis) in a number of segments. Each segment represents an *Age* category. For instance, five segments with their corresponding average values (grey rectangles) are presented in Figure 2.3. A regression line can then be seen as a smooth version of these averages. A consequence is that for each age value equal to, say x_0, the corresponding point on the regression line can be interpreted as the (approximate) average length given this X-value x_0. The average values in Figure 2.3 do not exactly lie on the regression line. However, it appears that for very large sample sizes, the average *Length* value for each *Age* value will exactly lie on the straight line (given that the relationship in the population is linear, see Section 2.5). Therefore, it does make sense to interpret the points on the regression line as (approximate) average Y values (average *Length*) for given values of X (*Age*). A very important point is that the dependent variable Y must be quantitative (continuous) because of the assumption of normal residuals (see Section 2.5).

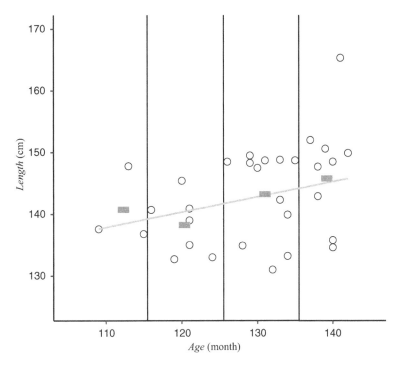

FIGURE 2.3
Regression line of *Length* on *Age*.

It should also be noted that a regression line is constructed in a completely different way, and the representation in terms of a graph of averages is just meant for our understanding how to interpret the regression line.

Actually, for each scatter plot, there is exactly one best (regression) line among all possible lines through the scatter plot.

> → *among all possible straight lines that we are able to draw through the scatter points, the regression line is the straight line that minimizes the sum of all (squared, vertical) deviations of observations from the regression line. This method of determining the regression line is denoted as the Ordinary Least Squares (OLS) method.*

Linear regression models, for which the regression parameters are estimated using the OLS method, will be called the OLS model in this book. From Figure 2.3, we also notice that there is a discrepancy between the points on the line and the observed values. The vertical deviation of each observation from the line is called residual. Due to variation of *Length* between the children (around the regression line) that may be due to intrinsic differences between the children (we will denote this as biological variation) and measurement error, there are residuals. Children may differ from each other due to genetic heritage or degree of sportiness (are you a swimmer or a weightlifter). Thus, for example not all 9-year-old children have the same body length, but there is some variation from the average *Length*. Moreover, body length cannot be measured without a measurement error. There is always a discrepancy between the actual body length and measured body length. Thus, the residual consists of a combination of biological variation and measurement error.

Recall that the points on the regression line are approximately equal to the average Y-values for very large sample sizes (equal to the population size). For a finite sample size (which will be the case in practice), these points will be called a predicted Y-value and is denoted as $\hat{Y}$ (Y-hat).

In statistics, we need to be able to quantify the predicted *Length* as a function of *Age* and so the residuals. Figure 2.4 shows a scheme of how the scatter points can be described (summarised) by a regression line plus residuals.

The asterisk in the figure represents an observation (with coordinate (X_i, Y_i)) for an arbitrary child *i*. The estimated regression line can be formalised by means of the following equation:

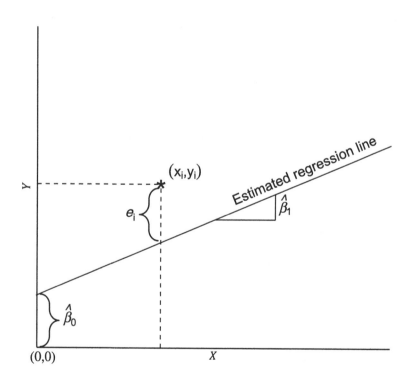

FIGURE 2.4
A schematic representation of the linear regression line.

$$\hat{Y}_i = b_0 + b_1 X_i \tag{2.4}$$

Note that a line is determined if both the intersection with the Y-axis (denoted by b_0 that will be obtained if $X = 0$ - see Equation 2.4 and Figure 2.4 denoted by $\hat{\beta}_0$) and the steepness (denoted by b_1) are known (in Figure 2.4 denoted by $\hat{\beta}_1$). For every child i, each observed value of X_i corresponds to one $\hat{Y}_i$ (of the same child i). This predicted value deviates from the observed Y_i according to

$$Y_i = \hat{Y}_i + e_i = b_0 + b_1 X_i + e_i, \tag{2.5}$$

with e_i as the residual.

→ *We also say that the estimated regression model is decomposed into a systematic (fixed) part $(b_0 + b_1 X_i)$ and a random part (e_i).*

The regression coefficients b_0 and b_1 are fixed, because they are the same for all the children. However, the residual e_i is random because it differs for each child (this is the reason why the residual has the subscript i but b_0 and b_1 do not). The regression model in Equation 2.5 estimates the population regression model

$$Y_i = \beta_0 + \beta_1 X_i + \varepsilon_i \tag{2.6}$$

where ε_i is the error (as a population statistics) and is generally unknown. To indicate that b_0 and b_1 are estimates of β_0 and β_1, respectively, in the following these estimators will be denoted by $\hat{\beta}_0$ (instead of b_0) and $\hat{\beta}_1$ (instead of b_1).

→ *It appears that the OLS estimators $\hat{\beta}_0$ and $\hat{\beta}_1$ are unbiased estimates of β_0 and β_1, respectively, with minimum variances[1] (Casella & Berger, 2002 p. 544). The standard deviation of an estimator $\hat{\beta}$ is called the standard error and measures the uncertainty of $\hat{\beta}$ as an estimator of the population parameter.*

The equation of the regression line is important for calculating the predicted *Length* for a given *Age*. The residuals are also important because they reflect how well the regression line summarises the scatter points by means of a straight line.

2.1.4 Interpretation of the Model Parameters

To interpret the results of a linear regression model, consider the results of a linear regression analysis of the nutrition study that is presented in Table 2.2.

---

[1] The minimum variance is actually among all linear unbiased estimators.

TABLE 2.2

Nutrition study: results of simple regression analysis of *Length* on *Age*

Parameter	Estimate	Std. Error	df	*t*-value	*p*-value	95% Confidence Interval Lower Bound	Upper Bound
Intercept	110.76	9.68	102.00	11.45	0.000	91.57	129.96
Age	0.25	0.07	102	3.31	0.001	0.10	0.39

Dependent Variable: *Length*.

Parameter	Estimate	Std. Error
Residual	45.27	6.34

Dependent Variable: *Length*.

What can we conclude from the output? The estimated regression line can be obtained from the output as

$$\widehat{Length} = 110.76 + 0.25 \times Age, \tag{2.7}$$

with the intercept $\hat{\beta}_0 = 110.76$ and slope $\hat{\beta}_1 = 0.25$. Thus, the intersection with the Y-axis is at *Length* = 110.76. This can also be calculated by filling in the value *Age* = 0 and then calculating the $\widehat{Length}$ using Equation 2.7.

→ *The effect of Age on Length is equal to the (approximate) average difference in Length between two (groups of) children whose ages differ from each other exactly one unit of measurement (here one month).*

For example, for two groups with *Age* = 120 and *Age* = 121, we can substitute *Age* in Equation 2.7 by 120 and 121 and subtract the corresponding predicted *Length* values from each other. Note that you can choose any two values of *Age*. That difference turns out to be

$Age = 121$ months: $\widehat{Length}_{(121)} = 110.76 + 0.25 \times 121 = 140.53$ cm
$Age = 120$ months: $\widehat{Length}_{(120)} = 110.76 + 0.25 \times 120 = 140.28$ cm
-- subtract
$$\widehat{Length}_{(121)} - \widehat{Length}_{(120)} = 0.25 \, cm.$$

This is exactly equal to the slope $\hat{\beta}_1$ of the regression line. Thus, the effect of *Age* on *Length* is about 0.25 cm per month, which is equivalent to 3 cm per year.

From Table 2.2, it can also be seen that the *p*-value for *Age* is 0.001, which is significant at 0.05 significance level, that is there is a significant effect of *Age* on *Length*. This implies that children who differ by 1 month will significantly differ on average by approximately 0.25 cm in body length (or about 3 cm per year on average). There is, however, considerable variation around the regression line (the standard deviation of the residuals is equal to SD(e) = 6.73 cm, which is the squared root of the estimated residual variance of 45.27 as presented in Table 2.2) and it reflects the biological variation (between-subjects variation) in *Length* for each *Age* plus measurement error. Note that 6.34 is the standard error of the estimated residual variance.

The design of the nutrition study is cross-sectional, and thus, we are not able to distinguish between measurement error and biological variation. As we will see later, a longitudinal design is able to make that distinction. Anyway, the results can be generalised to a population of children from which the children form the random sample. Nevertheless, it is sometimes difficult in practice to identify the real population. In this particular case, the population could be all urban/rural children coming from comparable districts.

When fitting a standard linear regression model, the underlying model (in the population) is of the form as indicated in Equation 2.6: $Y_i = \beta_0 + \beta_1 X_i + \varepsilon_i$. The error term ε_i is assumed to be independently and normally distributed with mean 0 and variance σ^2. We will come back to this point in detail in the next section. In practice, analysing a standard linear regression model with only one independent variable is too restrictive in most cases. Researchers may, for example, be interested whether the relationship differs between urban and rural children. In the next section, it will be highlighted that including an extra independent variable will be a better method than analysing standard linear regression models for urban and rural schools, separately.

2.2 Multiple Linear Regression (MLR) Modelling

2.2.1 Specification of a Multiple Linear Regression Model for the Nutrition Study

A relevant research question could be whether the relationship between *Age* and *Length* differs for urban and rural children. Recall that rural children on average are older but shorter than urban children (see Table 2.1), so a relevant question is whether it is still possible to compare urban and rural children.

In this section, we will demonstrate that despite the (average) *Age* difference between the two groups of children, a comparison can still be made by using the regression method. Two alternative approaches will be described and compared with each other:

a. An analysis for rural and urban schoolchildren, separately (subgroup analyses).
b. An analysis for rural and urban schoolchildren, together (joint analysis).

a) Sub-group (separate) analyses.

Using this approach, separate models are specified and analysed for rural and urban school children, separately. These models can be defined as follows:

$$Length_{i,rural} = \beta_0 + \beta_1 Age_i + \varepsilon_{1i} \text{ and}$$
$$Length_{i,urban} = \alpha_0 + \alpha_1 Age_i + \varepsilon_{2i}, \tag{2.8}$$

where α_0, α_1, β_0 and β_1 are population regression parameters to be estimated independently for urban and rural schools, respectively. In other words, the alphas are estimated using only the data of the urban schools, whereas the betas are estimated using only the data of the rural schools.

For the Nutrition study, estimates of the age effect α_1 (for the urban school) and β_1 (for the rural school) are of interest because these parameters reflect the strength of the relationship between *Age* and *Length* within each school type.

→ *If α_1 and β_1 are not equal, i.e., $\alpha_1 \neq \beta_1$, then the age effect depends on the school type. This phenomenon is called interaction. We then say that there is a School by Age interaction.*

Unfortunately, testing $\alpha_1 \neq \beta_1$ is not an easy job! Hence, subgroup analysis is not recommended in general. In addition, the sample size will decrease in subgroup analyses, which leads to a loss of power.

b) Joint analysis.

It is possible to study both groups using a single linear regression model. To accomplish this, the models describing two separate regression lines should be combined as follows:

$$Length_i = \beta_0 + \beta_1 School_i + \beta_2 Age_i + \beta_3 Age_i \times School_i + \varepsilon_i. \tag{2.9}$$

The terms *School* and *Age* are called the main terms or zero-order interaction terms and the product *Age* × *School* is called the first-order interaction or simply interaction term if there are no higher order interaction terms in the model.

Equation 2.9 is an example of a *multiple linear regression* model because the model contains more than one independent variable.

→ *A common rule in the specification of multiple linear regression models is to obey the hierarchical principle, i.e., if an interaction term (of certain order) is included in a regression model, all lower order interaction terms are also be included in the model.*

We will also follow this hierarchical principle throughout this book, except in one situation discussed in Chapter 5.

Because an interaction term is included in Equation 2.9, the hierarchical principle implies that the main terms *School* and *Age* must also be included in the model.

To see that the above model describes two different regression lines, we need to substitute the codes (which are 0 for a rural child and 1 for an urban child) for the variable *School* into the model.

$$Length_{i,rural} = \beta_0 + \beta_1 \times 0 + \beta_2 \times Age_i + \beta_3 \times \left(Age \times 0\right)_i + \varepsilon_i, \text{and}$$
$$Length_{i,urban} = \beta_0 + \beta_1 \times 1 + \beta_2 \times Age_i + \beta_3 \times \left(Age \times 1\right)_i + \varepsilon_i \tag{2.10}$$

Simplifying Equation 2.10 leads to

$$Length_{i,rural} = \beta_0 + \beta_2 \times Age_i + \varepsilon_i, \text{and}$$
$$Length_{i,urban} = \left(\beta_0 + \beta_1\right) + \left(\beta_2 + \beta_3\right) \times Age_i + \varepsilon_i \tag{2.11}$$

By substituting $(\beta_0 + \beta_1) = \alpha_0$ and $(\beta_2 + \beta_3) = \alpha_1$, it follows that the fixed part of Equations 2.11 and 2.8 is equivalent. Figure 2.5 also depicts the two regression lines. The difference in intercept between the two types of school is equal to β_1 (follows from Equation 2.11 by subtracting both equations from each other) and the difference in regression slopes is exactly equal to the interaction effect β_3. Clearly, if $\beta_3 \neq 0$, the two regression lines are not parallel, and the effect of *Age* is different for the two types of school.

→ *Consequently, the interaction effect β_3 in Equation 2.9 is the difference between the slopes of the regression lines for the urban and rural school.*

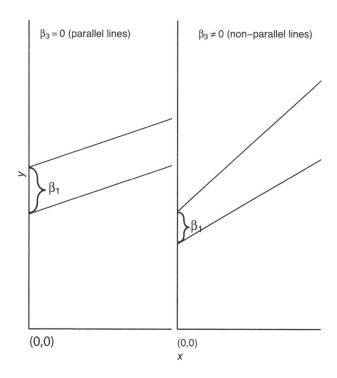

FIGURE 2.5
Visual interpretation of interaction.

Thus, the difference in *Age* effect between the two school types is determined by β_3.

2.2.2 Interpretation of the Model Parameters of the Nutrition Study

Regarding the first method mentioned in the previous section, the relationship between *Length* and *Age* was assessed by performing separate analyses. To interpret the corresponding results, we start with the scatterplot of the two types of *School* with their corresponding linear regression lines in Figure 2.6.

It can be seen that there is less growth among the rural school children, while a substantial growth is observed among the urban school children. We also notice considerable variability in *Length* for each *Age* category in both school types. Table 2.3 shows the estimated regression models in detail.

From the output of Table 2.3, it can be concluded that, among rural school children, no significant difference ($\alpha = 0.05$) in *Length* is observed for different ages (estimated slope $\hat{\beta}_1 = 0.112$ with *p*-value $= 0.380$), that is there is

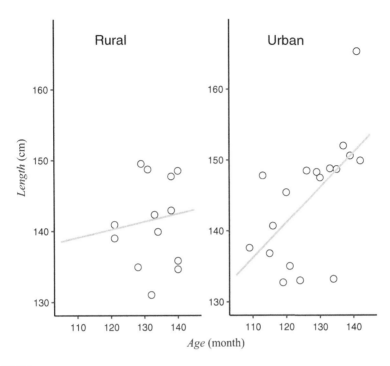

FIGURE 2.6
Scatter plot of *Length* against *Age* for rural (left panel) and urban (right panel) children.

TABLE 2.3
Nutrition Study: Result of Simple Regression Analysis of *Length* on *Age* per School Type

School type	Parameter	Estimate	Std. Error	df	*t*-value	*p*-value	95% Confidence Interval Lower Bound	Upper Bound
Rural	Intercept	126.78	16.83	54	7.53	0.000	93.04	160.52
	Age	0.112	0.13	54	0.89	0.380	−0.14	0.37
Urban	Intercept	81.60	12.33	46	6.62	0.000	56.78	106.42
	Age	0.496	0.10	46	5.07	0.000	0.30	0.69

Dependent Variable: *Length*.

School type	Parameter	Estimate	Std. Error
Rural	Residual	35.56	6.84
Urban	Residual	42.42	8.85

Dependent Variable: *Length*.

on average no clear growth in *Length* within the observed period of about 20 months (see Table 2.1). There is, however, a significant difference in body length among urban schoolchildren (estimated slope $\hat{\beta}_1 = 0.496$ with *p*-value = 0.000). Although the difference between the two school types is substantial and scientifically important, the separate analyses make it difficult to test for equality between the two regression slopes.

Suppose for the moment that the difference in regression slopes (0.11 vs. 0.50) is systematic, that is there is hardly any growth in the population of rural schoolchildren (approximately 1 cm per year), while the population of urban schoolchildren shows growth of about 6 cm per year. So, the difference in growth speed is about 5 cm per year.

> → *We can then conclude that there is an interaction effect between School and Age, i.e., the effect of Age on Length, depends on the type of school. It is estimated as 0.38 cm (which is equal to the difference between the two regression slopes) per month, or equivalently the estimated difference in growth speed is approximately 5 cm per year.*

To perform the analysis according to the second approach (the joint analysis), the multiple linear regression Equation 2.9 is specified. Table 2.4 presents the parameter estimates for Equation 2.9. The estimated models for each school type can be obtained by substituting the regression estimates into Equation 2.11 resulting in:

$$Length_{i,rural} = 126.78 + 0.112 \times Age_i + e_i, \text{ and}$$
$$Length_{i,urban} = 81.60 + 0.496 \times Age_i + e_i \tag{2.12}$$

TABLE 2.4

Nutrition Study: Result of Multiple Regression Analysis of *Length* on *Age* and *School*, and their Interaction.

Parameter	Estimate	Std. Error	df	*t*-value	*p*-value	95% Confidence Interval Lower Bound	Upper Bound
Intercept	126.78	17.56	100	7.22	0.000	91.94	161.62
Age	0.112	0.13	100	0.85	0.398	−0.15	0.37
School	−45.18	21.15	100	−2.14	0.035	−87.13	−3.23
Age × School	0.384	0.16	100	2.38	0.019	0.06	0.71

Dependent Variable: *Length*.

Parameter	Estimate	Std. Error
Residual	38.72	5.48

Dependent Variable: *Length*.

Note that the regression equations in 2.12 are exactly the same as obtained by performing the subgroup analysis (see Table 2.3).

Note, however, that the estimated residual variance is somewhere between the values obtained by the subgroup analyses, that is res.var.(rural) < res.var.(joint) < res.var.(urban) (35.56 < 38.72 < 42.42).

→ *In Equation 2.10 (joint analysis), the variance of the error terms for the two regression models are assumed to be equal, while in Equation 2.8 (subgroup analysis), the variances are not assumed to be equal.*

Without going into detail, it is possible to relax Equation 2.10 by allowing unequal error variances and analyse the model using a weighted least squares method (Draper & Smith, 1966, p. 108). Throughout this book, however, we will assume that the error variances for the two separate models are close to each other, so that we can assume equality.

From Table 2.4, it appears that the estimated difference between the slopes of the regression lines for the urban and rural school (which is equal to $\hat{\beta}_3 = 0.384$) is significantly different from zero (p-value = 0.019) and thus the effect of *Age* on *Length* for the rural schoolchildren differs significantly from that of the urban schoolchildren.

Interpretation of the other regression parameters depends on how *School* is coded and on the unit scale of *Age*. According to Table 2.4, the parameter β_2 of the regression model (Equation 2.9) is estimated by $\hat{\beta}_2 = 0.112$. This estimate can be interpreted as follows. By substituting the code 0 for school (as rural school = 0) in Equation 2.9, $\hat{\beta}_2$ can then be interpreted as the estimated *Age* effect for the rural school.

Anyway, we can conclude that there is hardly any growth for the rural schoolchildren, while there is a significant growth for the urban schoolchildren.

→ *It should be noted that, if School were coded differently, $\hat{\beta}_2$ would have a different interpretation with probably a different p-value (perhaps very small). This would not be the case if there is no interaction term. The interpretation and p-value of the interaction term is independent of how School is coded.*

According to Table 2.4, the parameter β_1 of the regression model (Equation 2.9) is estimated by $\hat{\beta}_1 = -45.18$. This estimate can be interpreted as follows. Fill in the code 0 for *Age* (new-born babies) in Equation 2.9. Consequently, $\hat{\beta}_1$ is the estimated *School* effect for new-born babies. It nevertheless makes no sense to interpret this estimate for the Nutrition study because no new-born babies were measured in the sample. The children from the rural area vary in *Age* between 121 and 140 months (see, Table 2.1). In general, it is not advisable to extrapolate the results outside the measured range of ages. The relationship between *Age* and *Length* may be very different outside this measured interval (perhaps nonlinear). It should also be noted that, because of the presence of an interaction term in the regression Equation 2.9, interpretation of the relationship between *Age* and *Length* should be performed per type of school. Furthermore, such analysis with a model specification as in Equation 2.9 with a discrete variable X does only make sense with a dichotomous X variable (here: *School*). Additional steps should be done when dealing with discrete X variables with more than two categories. This situation will be elaborated in Section 2.3. Finally, the results do not imply anything about the real growth of a certain child because the design of the study is cross-sectional. To address this point, we will need a longitudinal design where the measurements are repeated over time for each subject (child). Regression models for cross-sectional data are based on the assumption that all observations are independent. The OLS method can then be safely performed. Longitudinal data, however, are correlated as we will see later. Chapter 3 deals with the analysis of correlated data.

In the next section, we will analyse a (simulated) longitudinal case study with OLS demonstrating what can go wrong if the existing correlation between observations are neglected. We will see in the next section that conditioning on the subjects would solve the problem at the expense of generalisability to a larger group of subjects.

2.3 MLR Conditional on the Subjects

2.3.1 A Longitudinal Case Study: Violent-Behaviour Study

This section describes a longitudinal data set to study the relationship between violent behaviour (*Violent* as a quantitative variable; larger scores are associated with more violent behaviour) and alcohol consumption (*Alcohol* as a quantitative variable in units of glasses per week). The artificial data consist of five subjects each of which was measured five times between 1950 and 1958. Because not all subjects were measured in the same year, the design is called 'unbalanced design.' The dataset is generated such that *Violent* and *Alcohol* are positively correlated. Subjects drank more alcohol and showed more violent behaviour than subjects measured earlier in time. Actually, the positive relationship between *Alcohol* and *Violent* is due to a time effect: violent behaviour increases with time and people drank more over time.

The objective is to test the hypothesis that alcohol consumption is positively related to violent behaviour. Figure 2.7 shows the scatter plot of violent behaviour against alcohol consumption (in glasses) per week.

From this figure, it can be seen that there is a positive correlation between *Violent* and *Alcohol* ($r_{xy} = 0.82$), that is violent behaviour seems to increase as alcohol consumption increases. Furthermore, it can be seen from Table 2.5 with the results of the regression analysis on the 25 data points that the relationship is significant (*p*-value = 0.000, e.g., by evaluating the 95% confidence interval). Although intuitively very appealing, one could be tempted to conclude that there is a causal relationship, that is the more a person drinks, the more violent he/she will be.

Let us take a closer look at the scatter plot of Figure 2.7 and Table 2.5. The drawn line in Figure 2.7 is the regression function for all 25 data point, ignoring the fact that each of the five subjects was measured five times. The alcohol effect (i.e. the estimated slope) can be interpreted as the difference in average violent behaviour when alcohol consumption 'changes' one unit of measurement. Here, no distinction is made between measurements taken for one subject (within-subject repeated measurements) and measurements

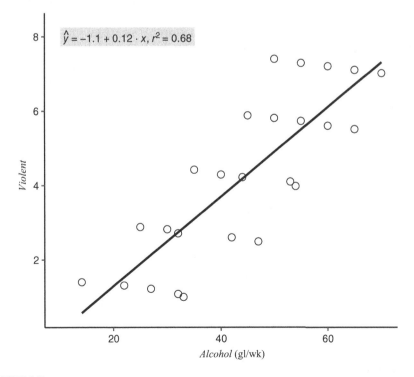

FIGURE 2.7
Scatter plot of violent behaviour against alcohol consumption in glasses per week.

TABLE 2.5

Violent Study: Results of Simple Linear Regression of *Violent* on Alcohol Consumption (*Alcohol*)

| | | | | | | 95% Confidence Interval | |
| | | Std. | | | | Lower | Upper |
Parameter	Estimate	Error	df	*t*-value	*p*-value	Bound	Bound
Intercept	−1.11	0.80	23	−1.39	0.178	−2.77	0.54
Alcohol	0.12	0.02	23	6.99	0.000	0.08	0.16

Dependent Variable: *Violent*.

taken for different subjects (between-subject measurements). Thus, the alcohol effect is based on a combination of the difference within subjects and the difference between subjects. In this particular study, the alcohol effect reflects to a large extent the difference in average violence behaviour between different subjects, that is the results are mostly based on a subject (or group) level.

However, the relationship does not say anything about the effect of alcohol consumption for a single subject, for example the effect on violent behaviour if a subject drinks more alcohol (by measuring alcohol consumption repeatedly over time). Figure 2.8 shows the scatter plot of *Violent* against *Alcohol* for each subject separately along with their corresponding estimated regression line.

It can be seen that although the overall correlation between *Alcohol* and *Violent* is positive, the correlation is negative for each subject separately.

→ *This phenomenon is called 'ecological fallacy'. The regression analysis as presented in Table 2.5 concerns an analysis based on aggregated data, which results do not necessarily hold at an individual level.*

For this problem, a multiple linear regression analysis is more suitable. The analysis should therefore be performed conditional on subjects, or equivalently we may say the regression analysis per subject. This is comparable to the Nutrition study, where *School* was included in the regression model, and the analysis was performed per school type.

In the present situation, if *Subject* is included in the model, the analysis can be performed per subject. There is, however, a problem that we need to tackle first. The variable *Subject* contains five categories (because the data contains five subjects). By considering *Subject* as a continuous independent variable and including it into the regression model directly (e.g. by using subject numbers 1–5 as scores), we implicitly assume that there is a linear relationship between *Violent* and *Subject*, conditional on (each value of) alcohol consumption. This does not make sense of course. One way to solve this issue is to express the variable *Subject* in terms of dummy variables. The next section provides a short intermezzo about dummy variables.

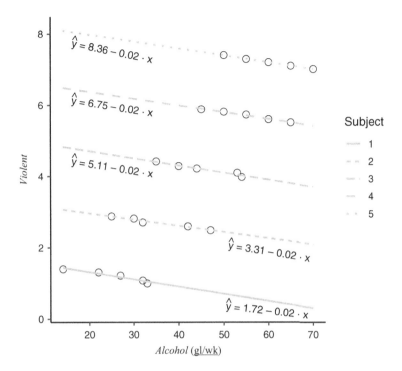

FIGURE 2.8
Regression lines of *Violent* against *Alcohol* for each subject, separately.

2.3.2 Intermezzo: Dummy Variables

To analyse the relationship between *Violent* and *Subject*, using a standard linear regression model as described above is incorrect because the variable *Subject* is of a nominal type with five categories having no rank ordering. The codes assigned to this variable are only meant to distinguish between these five subjects. Any assignment of a code to a subject is admissible. In Figure 2.9, for example, the *Violent* scores are plotted for five subjects. The subjects in the left part of the figure are assigned codes 1, 2, 3, 4, 5. With this code-combination, there is a perfect linear relationship between *Violent* and *Subject*. However, in the right part of the figure, the subjects are assigned different codes; 1, 5, 3, 4, 2, which is also admissible. As can be seen from the figure, with that code-combination, there is no perfect linear relationship between *Violent* and *Subject*. In fact, there is no linear relationship at all.

An alternative way to distinguish five subjects is by means of indicator variables. Here, we need to define four indicators $Dp1$, $Dp2$, $Dp3$ and $Dp4$. Each variable Dpi, $i = 1,\ldots, 4$ is an indicator for subject i. It assigns code 1 to Dpi if the measurement is from subject i and 0 otherwise. If all indicator variables are 0, the measurement is obviously from subject 5. A possible

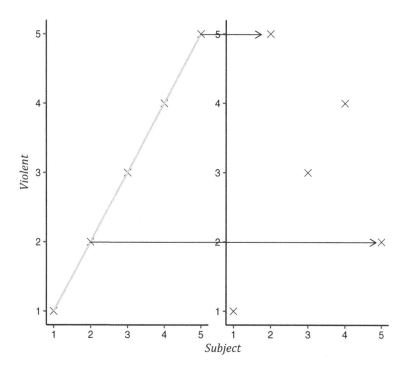

FIGURE 2.9
Relationship between *Violent* and *Subject* for different codings of *Subject*.

code-combination of indicator variables is shown in Figure 2.10 and Table 2.6.
Of course, other coding schemes are also admissible.

Such indicator variable is also called a dummy variable. In fact, four
dummy variables contain the same information as the nominal variable
Subject with five categories and should therefore be treated as one variable.
As a consequence, the multiple linear regression model of *Violent* on *Subject*
should be specified as follows:

$$Violent = \beta_0 + \beta_1 Dp1 + \beta_2 Dp2 + \beta_3 Dp3 + \beta_4 Dp4 + \varepsilon, \qquad (2.13)$$

where the dummy variables can be coded as given in Table 2.6. The expected
(population average and symbolised by E (.)) value of each subject can be
obtained by substituting the codes into

$$E(Violent) = \beta_0 + \beta_1 Dp1 + \beta_2 Dp2 + \beta_3 Dp3 + \beta_4 Dp4 \qquad (2.14)$$

Resulting in $E(Violent \mid subject\, i) = \beta_0 + \beta_i, i = 1, \ldots, 4$, and $E(Violent \mid subject\, 5) =$
β_0. The last subject (subject 5) is called the reference subject. All subjects are
being compared to this reference subject. The regression parameters can then
be interpreted as

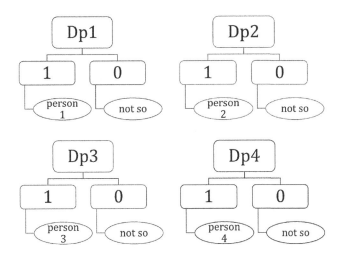

FIGURE 2.10
An alternative way to code a nominal variable.

TABLE 2.6

Violent Study: An Example of Dummy Coding for Subjects with the Last Subject as the Reference Category

Subjects	Dp1	Dp2	Dp3	Dp4
Subject 1	1	0	0	0
Subject 2	0	1	0	0
Subject 3	0	0	1	0
Subject 4	0	0	0	1
Subject 5	0	0	0	0

$$\beta_0 = E(Violent \mid subject\ 5)$$
$$\beta_i = E(Violent \mid subject\ i) - E(Violent \mid subject\ 5), \qquad (2.15)$$
$$i = 1,\dots,4$$

Thus, each regression parameter β_i can be interpreted as the difference between the expected violent behaviour of subject i and the reference subject 5. It should be noted that other coding schemes are also possible.

2.3.3 Analysis of Violent Data

For the ease of presentation, let us suppose (although it can also be tested) that all subjects show the same relationship with *Violent* (i.e. there is no interaction between *Dpi* and *Alcohol*). Therefore, the linear regression model can be specified as follows.

$$Violent_{ij} = \beta_0 + \beta_1 Alcohol_{ij} + \beta_2 Dp1 + \beta_3 Dp2 + \beta_4 Dp3 + \beta_5 Dp4 + \varepsilon_{ij} \qquad 2.16$$

where subscript ij indicates the j-th repeated measurement of subject i.

As the results in Table 2.7 indicate the relationship between *Violent* and *Alcohol* is significantly negative ($\hat{\beta}_1 = -0.02$ with p-value = 0.000). Furthermore, for each subject, the estimated regression line can be obtained from Table 2.7 by filling in the codes for the dummy variables as given in Table 2.6. This leads to

$$subject\,1 : E\left(Violent_{1j}\right) = 1.70 - 0.02 \times Alcohol_{1j},$$
$$subject\,2 : E\left(Violent_{2j}\right) = 3.38 - 0.02 \times Alcohol_{2j},$$
$$subject\,3 : E\left(Violent_{3j}\right) = 5.07 - 0.02 \times Alcohol_{3j}, \qquad (2.17)$$
$$subject\,4 : E\left(Violent_{4j}\right) = 6.76 - 0.02 \times Alcohol_{4j},$$
$$subject\,5 : E\left(Violent_{5j}\right) = 8.35 - 0.02 \times Alcohol_{5j}.$$

Because we assume no interaction between *Alcohol* and *Subject*, all regression lines are parallel, and the lines only differ from each other by the intercept estimates.

Unfortunately, the above results cannot be generalised to other (new) subjects because the model is conditioned on these five subjects (in the sample). Hence, all results apply only to these five subjects. In general, generalisation to a larger group of subjects is of greater interest. As we shall see in the next chapter, this is only possible if we use a different, more advanced regression technique that will be elaborated in extension.

TABLE 2.7

Violent Study: Results of a Multiple Linear Regression Model with Quantitative and Qualitative (nominal) Variables

Parameter	Estimate	Std. Error	df	t-value	p-value	95% Confidence Interval	
						Lower Bound	Upper Bound
Intercept	8.35	0.05	19	154.77	0.000	8.23	8.46
Alcohol	−0.02	0.00	19	−22.02	0.000	−0.02	−0.02
Dp1	−6.65	0.04	19	−185.06	0.000	−6.73	−6.58
Dp2	−4.97	0.03	19	−169.01	0.000	−5.04	−4.91
Dp3	−3.28	0.02	19	−137.61	0.000	−3.33	−3.23
Dp4	−1.59	0.02	19	−77.33	0.000	−1.63	−1.55

Dependent Variable: *Violent*. Subject 5 is the Reference Category.

2.4 Multiple Piecewise Linear Regression

2.4.1 A Comparative Longitudinal Study: Proximity Study

Sometimes, a straight line to describe the averages does not make sense or is not of interest. The proximity study (Brekelmans, Wubbels, and Créton, 1990) is such an example, wherein a total of 50 male (36) and female (14) teachers were evaluated on their interpersonal behaviour in the classroom. The degree of closeness between a teacher and his/her students was measured (variable *Proximity*; the degree of closeness increases with the scores). Each teacher was evaluated yearly for a period of three years, starting at baseline *Occasion* = 0, and proceeding with occasions = 1, 2 and 3 after one, two and three years, respectively. The dichotomous variable *Sex* had codes: *Sex* = 0 for male teachers and *Sex* = 1 for female teachers. The data also contained missing observations.

In this study, we are interested in the comparison between male and female teachers per occasion (time point) and/or how the average proximity scores change over time for male and female teachers separately. Figure 2.11 depicts the average proximity scores over time for male teachers (dotted line) and

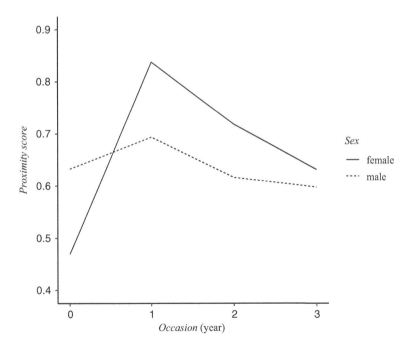

FIGURE 2.11
Mean profile of proximity scores versus measurement occasion for each *Gender*.

female teachers (solid line). Fitting a straight line through the averages of the female teachers, for example, would suggest that there is a constant change in the average proximity score over time. However, the curve for female teachers in Figure 2.11 suggests that there is an increase of average score after one year, and in the following years, the average score drops again towards the baseline level. Hence, describing the changes with straight lines would not answer the research question properly.

2.4.2 Specification of the Multiple Piecewise-Linear Regression Model

A way to describe how the average proximity scores change over time is to treat each measurement occasion as a separate group. The problem is comparable to an ANOVA problem, where four groups are compared and specified as a linear regression model (see, e.g., Field, 2018, Chap. 5). For the specification of the regression model, *Occasion* must be considered as nominal. Thus, *Occasion* must be expressed by three dummy variables. Furthermore, Figure 2.11 suggests that there is an *Occasion* by *Sex* interaction, that is the occasion effect differs for male teachers and female teachers. The model can be specified as follows (dropping the indices *i* and *j* for ease of presentation):

$$
\begin{aligned}
Proximity = \beta_0 &+ \beta_1 Occ_1 + \beta_2 Occ_2 + \beta_3 Occ_3 + \beta_4 Sex + \\
&\beta_5 Occ_1 \times Sex + \beta_6 Occ_2 \times Sex + \beta_7 Occ_3 \times Sex + \varepsilon.
\end{aligned}
\tag{2.18}
$$

In Equation 2.18, the baseline occasion is considered to be the reference time point, and Occ_1, Occ_2 and Occ_3 are the defined dummy variables with $Occ_j = 1$ if *Occasion* = j and 0 otherwise $j = 1, 2, 3$. Using the above model, each average proximity score at the population (denoted by expected score $E(\cdot)$) level can be expressed as a function of the regression parameters as follows:

$$
\begin{aligned}
E(Proximity_0) &= \beta_0 + \beta_4 Sex, \\
E(Proximity_1) &= \beta_0 + \beta_1 + \beta_4 Sex + \beta_5 Sex, \\
E(Proximity_2) &= \beta_0 + \beta_2 + \beta_4 Sex + \beta_6 Sex, \\
E(Proximity_3) &= \beta_0 + \beta_3 + \beta_4 Sex + \beta_7 Sex.
\end{aligned}
\tag{2.19}
$$

Now we can compare female and male teachers in terms of the regression parameters. Let us define the difference between male and female expected proximity scores by

$$
\Delta E(Proximity_j) = E(Proximity_j)_{female} - E(Proximity_j)_{male}, \text{ for } j = 0, 1, 2, 3,
$$

which denotes the sex difference in expected proximity score at occasion j. It then follows

$$\beta_4 = \Delta E\left(Proximity_0\right),$$
$$\beta_5 = \Delta E\left(Proximity_1\right) - \Delta E\left(Proximity_0\right),$$
$$\beta_6 = \Delta E\left(Proximity_2\right) - \Delta E\left(Proximity_0\right),$$
$$\beta_7 = \Delta E\left(Proximity_3\right) - \Delta E\left(Proximity_0\right).$$

(2.20)

In other words, β_4 is equal to the baseline difference (at occasion 0) in expected proximity score between female and male teachers. The parameter β_{j+4} is the difference in expected proximity score of female and male teachers at occasion j ($j = 1, 2, 3$) minus the expected baseline difference score, that is the parameters β_5, β_6, and β_7 express how the difference in true proximity score between male and female teachers changes from baseline to occasion 1, 2 and 3, respectively.

Using the observed data, the OLS method cannot be easily used to estimate the regression parameters of interest and the change in *Proximity* difference between female and male teachers. This is due to the existing correlation between the repeated measurements. In Chapter 3, we will introduce an advanced longitudinal regression method that enables us to evaluate this change over time and makes generalisations to a larger group of teachers possible.

2.5 Assumptions of the MLR with Uncorrelated Errors

Consider as an example the following multiple linear regression model with four independent variables.

$$Y_i = \beta_0 + \beta_1 X_1 + \beta_2 X_2 + \beta_3 X_3 + \beta_4 X_4 + \varepsilon_i,$$

(2.21)

where X_i, $i = 1, 2, 3, 4$ are independent variables. A proper fit of Equation 2.21, using the standard OLS method that produces unbiased estimates with the smallest standard error, requires certain assumptions.

These assumptions are:

1. Linearity
2. Normality
3. Independence

2.5.1 Linearity Assumption

→ *A general assumption about the functional relationship between the dependent variable Y and the independent variables X_1, ..., X_4 in Equation 2.21*

*is that there should be a linear relationship between Y and each (linear)
combination of the X variables. In particular, there should be a straight-line
relationship between Y and each of the X variables.*

If X is qualitative, then the dummy variables should be used and included in
the model. Note that the relationship between Y and a dichotomous X variable is already linear. Examples that we have described up to now are:

The violent-behaviour study. The variable *Violent* in Equation 2.15 is a quantitative dependent variable, the variable *Alcohol* is a quantitative independent variable, and the variable *Subject* is a qualitative independent variable.
According to the linearity assumption, there should be a straight-line relationship between *Violent* and *Alcohol* for each subject and the variable *Subject*
should be expressed as dummies.

The proximity study. The variable *Proximity* in Equation 2.18 is a quantitative dependent variable, the variable *Sex* is dichotomous, and the variable
Occasion is (treated as) qualitative. There is always a straight line between the
average *Proximity* score and any dichotomous variable and in particular *Sex*
for each occasion. The variable *Occasion* should be expressed as dummies.

In practice, relationships are not exactly linear except when the X variable is dichotomous. Still the linear regression model is relevant, because
in many cases, a description of the linear trend is sufficient, provided that
the observed relationship does not depart much from linearity or when the
amount of departure from linearity does not contain a content specific interpretation other than measurement error.

2.5.2 Normality Assumption

Recall that the observed dependent variable Y_i of Equation 2.21 can be decomposed into a fixed part containing all independent X variables and a random
part that consists of the error term ε_i. The normality assumption concerns the
random part ε_i, which is supposed to be normally distributed with mean zero
and variance σ^2. If the distribution is not normal, the OLS method will still
deliver unbiased estimates (Casella & Berger, 2002, p. 544), but the statistical
tests as implemented in almost all commercial computer programs are only
valid if the underlying distribution of errors is normal or if the sample size is
large enough. How large the sample size should be, depends, among others,
on the number of independent X variables in the model. The required sample
size will be smaller for models with two instead of 10 independent variables.
Sample size calculation is not a topic of this book and the reader is referred
to, for example, Altman (1991) and Bland (2000) for an introduction.

On the one hand, the observations $i = 1,\ldots, n$ form a random sample of
size n from a large group (population). Thus, the results of the analysis can
be generalised to this population. On the other hand, the model is specified
conditional on all X variables in the model. In particular, if X is categorical,

then the results are only valid for the categories of the *X* variable. Thus, if one of the X variables is *Subject*, then the results will only be valid for all subjects in the sample and cannot be generalised beyond these subjects.

Thus, if the normality assumption is fulfilled, we can perform hypotheses testing about the relationship between variables (such as effect of an independent variable *X* on the dependent variable *Y*) and generalise the results to the population from which the sample is considered to be randomly sampled. Below are two examples:

The Nutrition study. The index *i* in Equation 2.9 represents 104 urban/rural children in the sample. The results can be generalised to the population for which these children form a random sample. In practice, nevertheless, it can be difficult to identify what the population is because these 104 children were not really taken at random from a well-defined population of children.

The violent-behaviour study. The index *ij* in Equation 2.16 represents each of five repeated measurements of five subjects in the sample. These five subjects are fixed and therefore are not generalisable to a larger group of subjects.

2.5.3 The Independence Assumption

A very important assumption is the independence assumption, that is the error terms of Equation 2.21 are independent (uncorrelated or not associated with one another). Suppose that for a person, the deviation from a known regression line is observed. If the error terms are independent, there will be no information about the deviation from the regression line of any other arbitrary person. In particular, the errors as deviations from the regression line vary arbitrarily around the regression line. The correlation between two arbitrary errors ε_i and ε_j for $i \neq j$ with $i, j = 1,..., n$, (i.e., the correlation coefficient expressed in Equation 2.1) appears to be proportional to $r_{e_i,e_j} \approx \Sigma_{i \neq j} e_i \times e_j = 0$. In Chapter 4, we will see that we can test this independence assumption.

2.6 Issues of MLR for a Longitudinal Study

2.6.1 Implication of the Independence Assumption

Consider as an example from a longitudinal data design, where some distance measure (*Distance* in cm) on the age (*Age* in years) of one child is measured repeatedly over time. The corresponding model is:

$$Distance_j = \beta_0 + \beta_1 Age_j + \varepsilon_j, \tag{2.22}$$

with $j= 1,\ldots, 13$ repeated measurements.

Figure 2.12 shows the scatter plot of 13 repeated measurements of this child. The scatter plot displays a positive relationship between *Distance* and *Age*. However, the observations are not necessarily correlated. The positive relationship only refers to the positive functional relationship. Recall that the correlation between observations refers to the correlation between errors. In this situation, it is still possible that, despite the positive functional relationship, the errors are uncorrelated (independent). This situation may occur if, for example the child is measured repeatedly by a well calibrated computerised measurement device. The instrument does not have a memory, thus every time it measures the *Distance* of the child independently of what the outcome is in the previous year. Consequently, the repeated measurements are uncorrelated, and thus (suppose that the linearity and normality assumptions are met), the standard OLS method is valid and therefore leads to unbiased estimates.

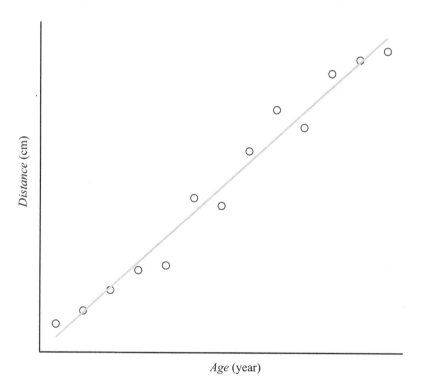

FIGURE 2.12
An illustration of longitudinal data for a child.

→ *If, on the other hand, the errors of the repeated measurements (given that there is only one subject) are correlated, then this correlation is called serial correlation. Apparently, $\Sigma_{i \neq j} e_i \times e_j \neq 0$ (see Equation 2.1), i.e., the deviation from the regression line does not vary arbitrarily.*

Next, consider the same study but now we have two children. Thus, Equation 2.22 applies for two children. Figure 2.13 shows the situation of two children measured repeatedly over time.

It is still possible to fit a regression line per child (conditional on each child) using the standard OLS method, but interest is usually on the regression line for the two children on average, that is the regression line through the scatter plot of averages. Note that for linear regression, this regression line is equal to the average regression line of the individual regression lines.

→ *In general, for linear regression models, we can say that the average (linear regression line) of all individual regression lines is equivalent to the linear regression line of the averages, which is sometimes handy for interpretation purpose.*

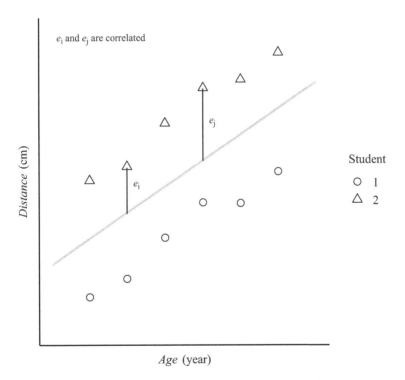

FIGURE 2.13
Average regression line of two children with no serial correlations.

Depending on the purpose of the study, the regression line as depicted in Figure 2.13 can be interpreted as the average of subject-specific (per child) regression lines or as the regression line through the scatter plot.

As can be seen from the figure, the repeated measurements of a child are all located above or all below the regression line, implying that the corresponding errors are correlated. In fact, the correlation is due to the fact that these two children differ from each other. The correlation between the repeated measurements of a child will be larger, as the two children differ more from each other, because the more the children differ from each other, the better we can predict whether two arbitrary chosen observations are coming from the same child or from different children. Thus, the repeated measurements are correlated, because the children differ. The case of more than two children is straightforward.

2.6.2 Shortcomings of the OLS Model

Apparently, in a longitudinal design with measurements of several subjects and several repeated measurements per subject, the data will be correlated. This correlation will be larger, as the subjects differ more from each other. This type of correlation is called the intraclass correlation (ICC). We will come back to this concept in detail in Chapter 3. The concept will then be formalised, leading to a formal definition of ICC.

The data of the Violent-behaviour study are obviously correlated. However, we have demonstrated that by including the subjects as a qualitative *X* variable, the relationship between *Violent* and *Alcohol* could be distinguished for each subject separately (conditional on each subject). In the absence of serial correlation, the assumption of independence is then fulfilled. Unfortunately, the results cannot be generalised to a larger group of subjects. To deal with that problem and the problem of correlated data in general, we need more advanced techniques than the OLS method. In the next chapter, multilevel analysis methods will be introduced that are suitable to handle correlated data.

2.7 Handling Missing Observations in Cross-Sectional Designs

Until now, we have assumed a complete data matrix in the regression analysis. However, the analysis can become complicated when there are missing observations, and hence, the data matrix is incomplete. As an example, consider the nutrition study wherein the aim is to predict student's length from his/her age. If *Age* is fully available for all children, but *Length* is only observed for some of them, we are confronted with the univariate missing data problem (see Figure 1.3). The natural question is therefore how to perform a regression analysis with missing observations for *Length*. Here, we distinguish two approaches proposed in the literature for handling missing data problems, namely, simple/ad hoc methods versus advanced/principled methods.

2.7.1 Simple Methods

2.7.1.1 Complete-Case Analysis

The simplest approach, known as *complete-case analysis (CCA)*, makes a complete matrix by removing incomplete cases. This means, cases for which some (or all) information is unavailable is totally removed from the dataset and therefore the remaining observations form a smaller matrix.

> → *A direct consequence of this method is a reduction in sample size, which potentially makes estimate of the regression coefficients less efficient. Moreover, the estimates can be biased when the data are non-MCAR (Little and Rubin, 2020, p. 49).*

Nevertheless, CCA should not be ruled out in general, as it may outperform sophisticated procedures in particular settings. White and Carlin (2010) showed its unique property through theoretical arguments and simulation studies in some scenarios. It should also be noted that many implementations of regression analysis in commercial software automatically adopt CCA by default.

2.7.1.2 Marginal Mean Substitution

Another simple method is the *marginal mean substitution*. Here, missing observations of each variable are replaced by the overall mean of the observed values from that variable (in the nutrition study, for instance, each missing observation of *Length* is filled in with the average of all observed values of *Length*). An advantage of this method is to produce a complete data matrix having the original size, for which the standard regression analysis can be performed.

> → *It is, however, discouraged because it tends to underestimate the standard error and to attenuate any correlations involving variables that are imputed.*

Nonetheless, it is a very attractive choice in randomised studies only when baseline variables have missing observations (Sullivan et al., 2018;

Kayembe et al., 2020). We defer the discussion to later chapters when we handle experimental studies.

2.7.1.3 Conditional Mean Substitution

An improved version of the previous approach is the *conditional mean substitution*, which imputes each missing observation by a conditional mean from the observed data. In the nutrition study, the missing observations of *Length* can therefore be replaced by the average of the observed *Length* values within each school type (i.e. separately for urban and rural schools). Note that the conditional means are generally constructed conditional on all variables in the dataset. In the nutrition study, for instance, *Age* and *School type* are used to obtain conditional means.

The conditional and marginal mean substitution methods generally lead to underestimated standard errors because the regression models do not supply uncertainty about the imputed values.

To obtain a better idea of substitution methods, suppose some children did not report their length (about 20%) in the nutrition study and hence have missing values in *Length*. The average of *Length* for the remaining children is 142.5 cm. This value thus should be substituted for each missing length value when the marginal mean substitution is used. For conditional mean substitution, the observed average of *Length* for rural and urban schools are, respectively, 141.6 and 143.7, and missing *Length* values are replaced by these numbers accordingly. Figure 2.14 shows a graphical representation of the substitution methods when the missing length values were substituted using the marginal mean (left) and the mean conditional on school type (right). In Figure 2.14, squared points are imputed values based on variants of mean substitution and circle points are observed values.

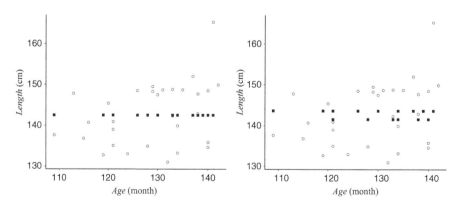

FIGURE 2.14
Scatter plot of *Length* against *Age* after mean substitution (left panel) and conditional mean substitution (right panel). Circles and squared points represent the observed and imputed Length values, respectively.

For the purpose of illustration, we also regressed *Length* on *Age* of children after applying each substitution method. In Table 2.8, their estimated regression slopes are compared with that of the complete data (before inducing missing values) as well as their root of mean squared error

$$\text{RMSE} = \sqrt{\frac{1}{n} \Sigma \left(y_i - \hat{y}_i \right)^2} \tag{2.23}$$

→ *The shortcomings of the marginal and conditional mean method are immediately obvious, because the association between Age and Length is weakened and the standard error of the regression slope is underestimated.*

2.7.1.4 Stochastic Regression Imputation

The conditional mean substitution method can be further improved by adding some noises to the conditional means, that is for each missing observation, a random noise will be added to its conditional mean to reflect uncertainty in the predicted value. This method is referred to as the *stochastic regression imputation* in the literature (Little and Rubin, 2020, p. 73). The method is superior to conditional mean substitution because the latter does not provide any variability around the imputed values. In order to elaborate on this point, suppose 10 students of the exact same age have missing values on *Length* in the nutrition study. The conditional mean substitution essentially replaces the same value for each missing entry (i.e. all 10 students get the same imputed length). However, this is a highly unrealistic situation in real life because students of the same age do not necessarily have exactly the same length. Hence, some variability should be added to the imputations to reflect realistic real-life situations.

We illustrate the stochastic regression imputation by focusing on the nutrition study, wherein the *Length* values are missing in about 20% of children. The first step is to obtain the predicted values (i.e. conditional means) for those children with missing length values. Thus, we fit a linear regression

TABLE 2.8

Nutrition study: estimates of the regression slope with standard error and root mean squared error (RMSE) from the linear regression model of *Length* on *Age*

Method	Regression Slope		
	Estimate	Stand. Error	RMSE
Complete data	0.25	0.08	6.73
Marginal mean substitution	0.13	0.07	5.98
Conditional mean substitution	0.19	0.07	6.01

model for *Length* that includes *Age* and school type as predictors. The estimated regression line from complete cases is

$$\widehat{Length} = 97.491 + 0.339 \times Age + 3.665 \times School \tag{2.24}$$

Now, the *Length* values are predicted from the above estimated model for each child with a missing value on *Length*. The next step is to add a random noise to each of these predicted values. With a linear regression model, the noise will be generated from a normal distribution with mean zero and variance equal to the estimated residual variance of the regression analysis (not shown). For categorical variables, the regression model gives predicted probabilities for each category, and then, the imputed values are drawn from those probabilities. Table 2.9 shows data of seven randomly selected students, where *Length* is imputed for students 1, 3, and 6. For instance, the predicted length of the first student is 138.5 = 97.491 + 0.339 × 121 + 3.665 × 0, and the noise is a random number generated from the aforementioned normal distribution. The results of the stochastic regression imputation are also depicted in Figure 2.15 for the entire study. Again, squared points represent imputed values from the stochastic regression imputation method.

→ *Although the stochastic regression method might appear appropriate at first glance, it is not advisable because it does not take into account the uncertainty due to missing data. In general, all single imputation methods ignore the fact that imputed values are estimated and not known, and therefore the standard errors are systematically underestimated resulting in narrower confidence intervals and smaller p-values (Carpenter and Kenward, 2008).*

TABLE 2.9

Nutrition Study: Stochastic Regression Imputation for a Selected Group of Students. School Type is Coded as 1 for Urban School and 0 for Rural School

Number	*Length* (cm)	*Age* (month)	School type	Predicted value	Random noise	Imputed value
1	NA	121	0	138.5	15.0	153.5
2	148.3	129	1	–	–	–
3	NA	124	1	143.2	−6.2	137.0
4	139.9	134	0	–	–	–
5	139.0	121	0	–	–	–
6	NA	116	1	140.5	−8.9	131.6
7	148.5	140	0	–	–	–

'NA' indicates a missing value.

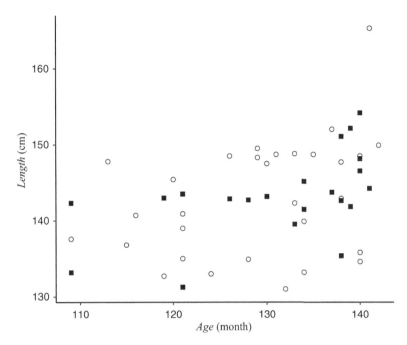

FIGURE 2.15
Scatter plot of *Length* against *Age* after stochastic regression imputation. Circles and squared points represent the observed and imputed Length values, respectively.

The other simple methods such as missing indicator method and last observation carried forward will be discussed in the next chapter.

2.7.2 Advanced Methods: Direct Likelihood and Multiple Imputation

We review two general advanced methods of dealing with missing data in this section. The direct likelihood (DL) approach is based on maximum likelihood estimation. In short, the likelihood function is first defined, which is a loose translation of the probability function that describes the contribution of each subject in the estimation process. The function then seeks plausible values (known as estimates) for the parameters that together with the data deliver the highest value for the likelihood function. The parameter value at which the likelihood function attains its maximum for the data at hand is called the maximum likelihood estimate (MLE). For incomplete data, missing observations do not contribute to the likelihood function, so the *observed likelihood function* is used to obtain the MLEs (for more detailed material, see, for example, Little and Rubin, 2020, Chap. 6). The DL method is also known as full information maximum likelihood (FIML) in the literature. Because a closed-form solution for the observed data likelihood cannot be found in general, iterative methods such as the expectation-maximization (EM) algorithm or its extensions can be applied. This is particularly useful when the pattern of missing data is arbitrary (or non-monotone – see Chapter 1). Enders (2010) provided a nice overview of the EM algorithm using an illustrative example in the presence of missing data.

In cross-sectional studies, the DL approach reduces to CCA when missing observations are confined to the dependent variable only (under the ignorable missingness mechanism). This is attributed to the fact that the method only uses complete cases because subjects with missing values in the dependent variable do not contribute to the likelihood function (as their values are missing). In contrast, if the independent variable has missing observations in addition to the dependent variable, the DL approach does not reduce to CCA because for subjects with missing observations in the independent variables, their corresponding dependent variable (which is observed) still contributes to the likelihood function. Note that the DL approach is not readily available in routine statistical procedures so that iterative optimisation algorithms such as the EM algorithm are required to handle the problem of missing observations in the independent variables.

Multiple imputation (MI – Rubin, 1987) is another advanced method, which refers to a procedure that replaces each missing observation in a data matrix with two or more acceptable values, that is the so-called imputed values, thus resulting in multiple completed data sets. Multiple imputation procedure consists of three separate phases, namely, Imputation phase, analysis phase, and pooling phase. In Figure 2.16, we start with an incomplete matrix, wherein some values are missing (blank parts). The observed values are represented by a crossed pattern in Figure 2.16. The imputation phase repeatedly fills in these blanks (depicted by different grey colours) for each missing observation using a probabilistic model, known as the predictive distribution. The analysis phase (circles in the middle) performs standard complete-data procedures to obtain the parameters of interest (i.e. the regression coefficients) for each imputed data set. Finally, the pooling phase combines the results from the separate analyses to form a single inference. That is, the pooling phase reports a set of parameters as the final estimates of interest, for instance, the regression coefficients. In short, the pooling phase consists of two parts:

1. Estimates of a parameter across imputations are averaged to form the final estimate.
2. The standard error of the final estimate is obtained by combining both between and within-imputation variances.

MI can be viewed as an upgraded version of the stochastic regression imputation. Suppose, for the moment, that the missing *Length* values in the

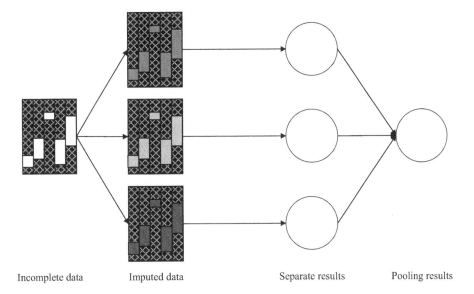

FIGURE 2.16
A schematic representation of multiple imputation procedure when each missing value is imputed three times (different colors show different imputations).

nutrition study are singly imputed using the stochastic regression imputation approach. The left panel of Figure 2.17 shows the fitted regression of children's length on their age in the completed data. If this regression line is the final estimate of the model to interpret the association between *Length* and *Age*, the imputed values of *Length* (squared points in the left panel of Figure 2.17) are then treated as if they are known. However, these values are

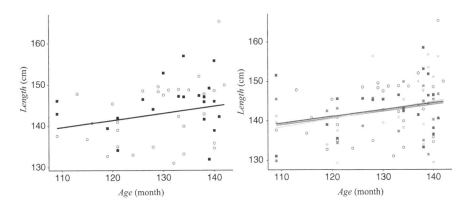

FIGURE 2.17
Scatter plot of the completed data in the Nutrition study when the missing *Length* values are singly (left panel) or multiply (right panel) imputed.

just educated guesses and not actually observed (unlike the known observed length values – circle points in Figure 2.17). Hence, it has been proposed to repeat, in parallel, the imputation process multiple times and obtain multiple fitted regression lines in order to take uncertainty due to missing data into account (see, among others, Rubin, 1987; Carpenter and Kenward, 2013; van Buuren, 2018; Raghunathan, Berglund and Solenberger, 2018).

→ *It should be noted that creation of multiple completed data sets from stochastic regression imputation method does not fully propagate uncertainty due to missing data because only one model is entertained.*

This means that an identical estimated model (i.e. Equation 2.24 in the nutrition study) is used to create each imputed data set, and therefore, there is no variability between imputed data sets, which, in turn, invalidates any inference obtained from MI [see, e.g. Murray (2018) for a more detailed discussion].

→ *Hence, an appropriate MI procedure must essentially account for two sources of uncertainty: (i) sampling variability assuming the reasons for missingness are known (within imputation variability), and (ii) variability due to uncertainty about the reasons for missingness (between imputation variability).*

The right panel of Figure 2.17 shows the MI procedure applied to the nutrition study, wherein every missing *Length* value is imputed three times (depicted by different grey colours). Each of three completed data sets is then used to estimate the corresponding regression line (shown by different grey colours as well).

2.7.3 Generating Imputations

Imputation methods can be classified into three classes in general: Joint modelling approach assuming multivariate normality or its variants (Schafer, 1997; Carpenter and Kenward, 2013), multiple imputation by chained equations (MICE), also known as sequential regression imputation or fully conditional specification, which defines a set of conditional models for imputation (van Buuren et al., 1999; Raghunathan et al., 2001; White et al., 2011), and hot deck imputation that matches missing values to the observed values (Reilly, 1993; Andridge and little, 2010). Our focus is on MICE because it is simple, flexible, and widely available. However, the theoretical properties of MICE are not well understood. Recently, Liu et al. (2014), Hughes et al. (2014), and Zhu and Raghunathan (2015) have shown that MICE matches the joint modelling approach in various situations in cross-sectional designs. Furthermore, experience with MICE through simulations and real applications does not seem to suggest that lack of full theoretical justification should be of great concern in practice.

MICE is based on a variable-by-variable principle, wherein the first incomplete variable is regressed on the other variables in order to obtain imputations for that variable. The same rule applies to the other incomplete variables in turn until all variables with missing values are imputed. The whole cycle is consecutively repeated until apparent convergence is reached, which is usually achieved after a few iterations (about 10 or 20). The whole process is repeated M times to produce M completed (imputed) data sets. In order to illustrate the procedure, suppose *Length* and *Age* of some children in the nutrition study are missing. The MICE approach implies that the child's length is regressed on his/her age to impute missing *Length* values. Likewise, the child's age is regressed on his/her length to impute missing *Age* values. In mathematical formulation, the following regression models known as the *imputation model* are consecutively used for imputation:

$$\begin{aligned} Length &= \alpha_0 + \alpha_1 Age + \epsilon_{Length} \\ Age &= \gamma_0 + \gamma_1 Length + \epsilon_{Age} \end{aligned} \quad (2.25)$$

A challenge in MI is the specification of the imputation model, because a mis-specified imputation model can lead to biased parameter estimates in the main analysis (Collins et al., 2001; Moons et al., 2006; Schafer, 2003). As a general rule, the imputation model should be as general as possible in order to (1) preserve all associations among variables, and (2) make the MAR assumption more plausible. In the following, we provide some practical guidelines to avoid bias and gain precision when selecting variables for the imputation models (see, van Buuren, 2018, pp. 163–164 for an elaborate discussion):

1. All dependent and independent variables that are in the main analysis (i.e. the substantive model) should be added to the imputation model.
2. All variables that are predictors of the incomplete variable should also be included in the imputation model.
3. Auxiliary variables that can explain a considerable amount of variation in the incomplete variable are good sources to improve imputations.
4. The imputation model should also reflect any other association among variables such as interactions, nonlinearity, and a multilevel structure.

Hence, the general recommendation is to specify a rich imputation model using all available information (e.g. secondary and auxiliary variables and proxies), and subsequently impute missing observations by taking into account uncertainty due to both missing values and imputation model. In the following section, we provide an application of MI in the nutrition study. For the sake of presentation, we have made randomly about 20% of *Length* values missing for some children, while *Age* and school type are fully observed for all children. In Section 2.2.1, we have seen that the analysis model can be defined as follows:

$$Length_i = \beta_0 + \beta_1 School_i + \beta_2 Age_i + \beta_3 Age_i \times School_i + \varepsilon_i \qquad (2.26)$$

Drawing good imputations for missing *Length* values requires carefully specifying the imputation model. Here, both child's age and school type should be included in the imputation model (as independent variables in the analysis model). Moreover, the analysis model involves the product *Age* × *School* (i.e. the interaction between *Age* and *School*) so that the imputation model should preserve such association when the imputations are drawn. Therefore, the following imputation model should be defined:

$$Length_i = \alpha_0 + \alpha_1 School_i + \alpha_2 Age_i + \alpha_3 Age_i \times School_i + \varepsilon_i \qquad (2.27)$$

The standard multiple imputation procedure does not include the product term in the imputation model by default. We thus create the interaction term separately and include it in the imputation model as a separate independent variable. Table 2.10 shows the estimated regression coefficients for incomplete data, the first imputed dataset, and the pooled results after generating 100 imputed datasets.

It should be noted that if an independent variable (or covariate) has missing observations, a suitable imputation model must also be defined for this variable and subsequently added to the above set (see, e.g. van Buuren 2018, Chap. 4). As an example, suppose school type (coded as a binary variable) is not reported for some children. The imputation model for *School* can thus be a logistic regression of the probability of attending, for example the rural school conditional on *Age* and the dependent variable *Length*.

Here, a subtle issue is how to deal with the interaction *Age* × *School* in the imputation Equation 2.27 because this variable cannot be computed when *School* is not fully observed. A simple solution to this is to first impute missing observations in *School* and then calculate the interaction *Age* × *School*. The calculated interaction, known as the 'passively imputed' variable, is afterwards used to impute missing observations of *Length* in Equation 2.27. For a thorough discussion, see, among others, Seaman, Bartlett, and White (2012), Bartlett et al. (2015), and van Buuren (2018, Chap. 6.4).

An important practical question concerns the choice of the number of M imputations. Classical texts suggest that few numbers of imputed data sets (three to five imputations) are adequate. However, other research shows that a large number of imputations might be beneficial in practice (White et al., 2011).

TABLE 2.10

Estimates of the Regression Coefficients for Incomplete Data, First Imputation, and the Pooled Estimates from 100 Imputations

| | | | | | | 95.0% Confidence Interval for B | |
| | | | | | | --- | --- |
	Parameter	Estimate	Std. Error	*t*-value	*p*-value	Lower Bound	Upper Bound
Incomplete data	Intercept	134.68	19.43	6.93	0.000	96.01	173.36
	School	−47.16	24.18	−1.95	0.055	−95.29	0.97
	Age	0.05	0.15	0.36	0.724	−0.24	0.34
	Age × School	0.40	0.19	2.14	0.036	0.03	0.77
Imputation 1	Intercept	179.06	28.45	6.29	0.000	122.61	235.51
	School	−84.18	34.26	−2.46	0.016	−152.14	−16.22
	Age	−0.30	0.21	−1.38	0.171	−0.72	0.13
	Age × School	0.69	0.26	2.62	0.010	0.17	1.21
Pooled	Intercept	138.13	22.92	6.03	0.000	92.98	183.29
	School	−53.13	26.61	−2.00	0.047	−105.47	−0.78
	Age	0.02	0.17	0.14	0.889	−0.32	0.37
	Age × School	0.45	0.20	2.18	0.030	0.04	0.85

→ *In theory, MI is as efficient as the direct likelihood approach (under MAR) when the number of imputations is infinite (Rubin, 1987).*

Nevertheless, it is practically infeasible or even impossible to produce a gigantic number of imputations in real-life applications simply because of computation and storage limitations. A relevant question is therefore how small should the number of imputations be to obtain satisfactory results? The answer to this question depends on several factors among which statistical efficiency of point estimates (i.e. increased in variance of point estimates based on M imputations relative to that based on an infinite number of imputations) and coverage probability of interval estimates are of main importance. Rubin (1987, pp. 114 and 115) focused on these aspects and argued that a small number of imputations ($M < 10$) is sufficient for most applications. If we are interested in other quantities such as *p*-value and statistical power, however, a larger number of imputations is usually needed. It has been shown that the number of imputations should be somewhere in the range of 20–200 imputations in such cases (Royston, 2004; Graham et al., 2007; White et al., 2011; Carpenter and Kenward, 2013; Von Hippel, 2020).

Fraction of missing information (FMI) is another factor that plays an important role when deciding on the number of imputations. For a parameter, it can be interpreted as the fraction of the total variance (both between and within-imputation variance) that is attributable to between-imputation variance when the number of imputations is infinite. As discussed in the study by White et al. (2011) and references therein, the following rule seems satisfactory in practice.

→ *As a rule of thumb, the number of imputations should be close to the percentage of incomplete cases as long as the FMI is less than the percentage of incomplete cases.*

A difficulty with the above rule is that the FMI is unknown, although it can be estimated. Moreover, the FMI can be different for different parameters. We therefore make the following lines of recommendations for practice: It is worth to start with a small number of imputations ($M = 5$). On the one hand, if there is a clear-cut inference and main interest lies in the point estimates (e.g. regression coefficients), there may not be much gain to go beyond five imputations. On the other hand, if there is an interest in some quantities, for instance, variance components in multilevel models, or we want to be confident about reproducibility of the results, it may be appropriate to go for a large number of imputations (e.g. $M = 100$). Likewise, one may use the estimate of FMI from the five imputed data sets and use the rule of thumb to set the minimum required number of imputations.

2.7.4 Comparison of Methods

We have reviewed several simple and advanced methods for handling missing observations in the previous sections. In this section, we compare these methods with respect to their validity in cross-sectional studies. The comparison is made for different scenarios.

2.7.4.1 *Partly Missing Dependent Variable Related to Fully Observed Independent Variables*

We first focus on situations in which missing observations are confined to the dependent variable only.

→ *Here, complete-case analysis (CCA) is valid and produces unbiased estimates as long as the reasons for missingness, after conditioning on independent variables, are unrelated to the dependent variable. This is because the remaining sub-sample (i.e., complete cases) can be considered as a random sample from the original population.*

In the nutrition study, suppose that all children who are older than 130 months have missing values on *Length*. This means both *Age* and *Length* values are observed for children younger than 130 months (i.e. the filled circles in the left panel of Figure 2.18), while only the *Age* values are observed for children older than 130 months (i.e. the *Length* values are missing for those children).

The left panel of Figure 2.18 also shows the fitted regression lines of *Length* on *Age* (i) when the data are complete (the grey line in the left panel of Figure 2.18) and (ii) when children older than 130 months have missing observations on *Length* (the black line in the left panel of Figure 2.18).

The figure elucidates that CCA is an appropriate choice, as the regression line can unbiasedly be estimated. In this case, CCA is fully efficient because children older than 130 months provide no information about the regression coefficients.

2.7.4.2 Partly Missing Dependent Variable Related to the Dependent Variable Itself

→ *When the reasons for missing observations in the dependent variable are related to the dependent variable itself in addition to the other independent variables, if any, CCA is invalid and produces biased regression coefficients because the remaining subsample is no longer a random sample from the original population.*

In the nutrition study, suppose now all children whose body length is higher than 140 cm have missing observations on *Length* (the unfilled circles in the right panel of Figure 2.18). This implies that only *Age* is available for children smaller or equal 140 cm. Figure 2.18 (right panel) compares the estimated regression lines for complete data (i.e., before introducing missing observations in *Length* – the grey line) and complete cases (the black line) and visibly

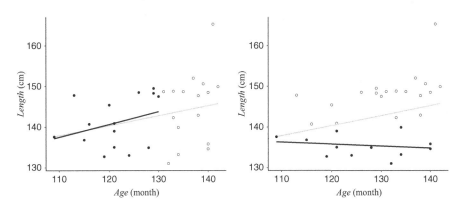

FIGURE 2.18
Regression lines of *Length* on *Age* when children older than 130 months are removed (left panel) and when children taller than 140 cm are removed (right panel). Gray lines are the estimated regression line before introducing missing observations on *Length*.

shows that the slope of the regression line is badly underestimated when complete cases are only used. It should be emphasised that the standard implementation of MI and DL approaches do not solve this problem and the estimated regression coefficients will be biased as well.

If there is a strong conviction that the missingness on the dependent variable depends on the dependent variable itself, even after conditioning on the other independent variables, the current recommendation is to repeat the analysis under different modelling assumptions and to investigate how sensitive the results are with respect to these assumptions. Later, we will discuss how to conduct sensitivity analysis in the presence of MNAR observations in the dependent variable in longitudinal settings.

2.7.4.3 Partly Missing Independent Variable Unrelated to Fully Observed Dependent Variable

Suppose the dependent variable is fully observed but the independent variables are partially observed. In the nutrition study, for example, age and school type might be missing for some children, while *Length* is fully observed for all children. Similar to the missing observations in the dependent variable (depicted in the left panel of Figure 2.18), CCA is valid and produces unbiased estimates as long as the reasons for missingness in the independent variables are unrelated to the dependent variable. This includes missingness mechanisms that are MCAR (a random missingness), MAR (missingness depends on the other independent variables but not the dependent variable), and MNAR (missingness depends on the independent variable itself).

2.7.4.4 Partly Missing Independent Variable Related to the Dependent Variable

However, CCA is not appropriate if the missingness in the independent variables depends on the dependent variable, even after conditioning on the other independent variables. As an example, suppose there are more children with missing *Age* values among taller children (similar to the right panel of Figure 2.18). This means the chance of not observing the child's age is bigger for taller children than the shorter ones. This is therefore an example that the missingness in the independent variable depends on the dependent variable. CCA is inappropriate here, as it produces biased estimates, but advanced methods such as direct likelihood or multiple imputation can produce unbiased estimates and hence are appropriate.

In order to have a clear picture about the performance of these different methods, we conducted a limited Monte Carlo simulation that resembles the nutrition study. First, hypothetical *Length* and *Age* values were generated for 104 children using a regression of *Length* on *Age* (on original data). Second, missing values were introduced in the dependent variable (*Length*), in the independent variable (*Age*), or in both variables. The missingness mechanism depended on the dependent variable, the independent variable, or on both, which resulted in several scenarios of incomplete data (see Table 2.11). We have set the percentage of missing data in each scenario to 50%. Third, the slope of the regression of *Length* on *Age* and its corresponding standard error were then estimated for complete, incomplete, and completed data sets. For multiple imputations, five completed data sets were created, and the corresponding results were pooled subsequently. Table 2.11 shows the estimated slope coefficient for each method after averaging across 500 replications.

Except for the mean substitution (or mean imputation), the other approaches, namely, CCA, MI and DL, produce unbiased estimate of the regression coefficient when the missing observations are totally random. If the missing data mechanism depends on the independent variables but not on the dependent variable, then CCA, MI and also DL considered to be valid. Note that this also includes an MNAR mechanism when the missingness on the independent variable depends on the independent variable itself. In contrast, if the missing data mechanism depends on the dependent variable, CCA is invalid due to a biased estimate of the regression coefficient. The advanced methods MI and DL were then only valid when the missing observations are confined to the independent variables (scenario 5 of Table 2.11). For the other scenarios where all methods failed (scenarios 2 and 8 of Table 2.11), the missing data mechanism should also be included in the analysis as a result of an MNAR mechanism, and hence, sensitivity analysis should be preferred.

2.7.5 Conclusion and Recommendations

In sum, we may conclude that CCA is a valid approach in cross-sectional studies (for continuous dependent variables) as long as the dependent variable is not the cause of missing observations. It should however be noted that CCA implies a smaller sample, which, in turn, has influences on the statistical power of a test and on efficiency. When missing observations are limited to the independent variables and the missing data mechanism depends on the dependent variable, the standard implementation of MI and DL approaches is recommended. Finally, if the dependent variable is the cause of missing observations (in the dependent variable or both), such advanced methods should be tailored to account for the MNAR mechanism, and sensitivity analysis would be the standard procedure. From a practical point of view, this means that the corresponding analysis should be repeated under different assumptions that are made about the missing data mechanism. For imputation of missing observations, this can be done, for example, by making imputations under different imputation models and then comparing the results (e.g. the regression slope) with respect to these assumptions. This point will be discussed through examples in Chapters 5–7.

TABLE 2.11

Average Estimate of the Regression Slope with its Standard Error (in parentheses) from 500 Replications of the Simulation Study

Scenario	Missing Values in	Missingness Depends on	Methods Complete Data	CCA	MS	MI	DL
1	Dependent variable	None	0.25 (0.075)	0.25 (0.108)	**0.12** (0.054)	0.26 (0.114)	0.25 (0.107)
2		Dependent variable	0.25 (0.075)	**0.12** (0.079)	**0.06** (0.038)	**0.13** (0.083)	**0.12** (0.080)
3		Independent variable	0.25 (0.075)	0.26 (0.162)	**0.06** (0.054)	0.26 (0.169)	0.26 (0.162)
4	Independent variable	None	0.25 (0.075)	0.24 (0.108)	0.24 (0.111)	0.24 (0.104)	0.25 (0.106)
5		Dependent variable	0.25 (0.074)	**0.13** (0.079)	**0.13** (0.115)	0.23 (0.131)	0.24 (0.133)
6		Independent variable	0.24 (0.074)	0.25 (0.159)	0.25 (0.166)	0.25 (0.167)	0.26 (0.166)
7	Dependent variable & independent variable	None	0.24 (0.075)	0.24 (0.101)	**0.18** (0.077)	0.23 (0.100)	0.24 (0.099)
8		Dependent variable	0.24 (0.075)	**0.12** (0.076)	**0.10** (0.058)	**0.14** (0.087)	**0.14** (0.086)
9		Independent variable	0.25 (0.075)	0.26 (0.163)	**0.20** (0.108)	0.25 (0.166)	0.27 (0.163)

The methods are complete data (before introducing missing observations), complete-case analysis (CCA), mean substitution (MS), multiple imputation (MI), and direct likelihood (DL).

Table 2.12 summarises our findings by highlighting which method is the minimum suitable method to handle missing observations in cross-sectional studies with a continuous dependent variable. In short, we recommend using the DL or MI approach as the starting point for conducting sensitivity analysis when the missing data mechanism depends on the dependent variable. This practice can be optional when the missingness is confined to the independent variables only (second row), as the DL and MI approaches are valid if the MAR mechanism is correct, that is, the missingness of independent variable(s) depends on the dependent variable. We emphasise that the MAR assumption cannot be verified by the data at hand, and therefore, validity of these methods highly depends on this assumption.

In practice, sometimes, it is hard to realise what precisely causes missing observations. It might even be harder for researchers to find out whether the dependent variable is related to the missing observations or not. A relatively simple way is to compare the distribution of a fully observed variable, say variable A, in the observed and missing part of the data. If these two distributions (obtained, respectively, from the observed and missing part) are

TABLE 2.12

The Suitable Methods to Handle Missing Data in Cross-Sectional Studies. Methods Are Complete-Case Analysis (CCA), Direct Likelihood (DL), Multiple Imputation (MI), and Sensitivity Analysis (SA)

Missing Data		Mechanism	
		Depend on Dependent Variable	Do Not Depend on Dependent Variable
location	Dependent variable	DL/MI + SA	CCA
	Independent variable(s)	DL/MI (+ SA)	CCA
	Both	DL/MI + SA	CCA

relatively comparable, it can indicate that the variable A is not a predictor of missingness and thus unrelated to the missing data mechanism. As an example, suppose some values of *Age* in the nutrition study are missing, while the length of children is fully observed. We therefore need to compare the distribution of *Length* for children with and without *Age*. If it turns out that both distributions are more or less equivalent in the sense of having similar means, variances and shapes, it might be concluded that the missingness on *Age* is not related to *Length* and therefore CCA can produce unbiased results.

Although this approach is straightforward, it can become complicated if more than one variable has missing observations. Imagine in addition to missing observations in *Age*, some other children have unmeasured *Length* values. It will then be quite difficult to compare the distribution of any of these variables with the other one (simply because of missing observations on that variable). In such situations, we recommend using various missing data methods and performing the same analysis in the incomplete and completed (i.e. imputed) data sets. By comparing similarities and dissimilarities of the results from different methods, we may be able to understand, to some extent, the possible causes of missing data. For instance, if all methods perform similarly, it might indicate that the missing data mechanism is presumably MCAR, though the possibility of the MNAR mechanism cannot be ruled out. In contrast, if MI and DL produce equivalent results but different from CCA, it implies that the missing data mechanism is not MCAR. Hence, the missingness is not purely random and the sensitivity analysis is preferred.

2.8 Assignments

2.8.1 Assignment

Consider a linear regression model $Y = \beta_0 + \beta_1 X + \varepsilon$. The mean of X is 42, the mean of Y is 72.

a. If $r_{XY} = 0$ (and thus cov $(X, Y) = 0$), then what is the predicted value of Y when $X = 30$?

b. Suppose $\hat{\beta}_0 = 1$. Determine $\hat{\beta}_0$

2.8.2 Assignment

Consider the following SPSS system file: 'Nutrition.sav'.

The study concerns a random sample of 104 schoolchildren, divided into 56 rural (schools) and 48 urban (schools) children. The variable *Age* is measured in months and *Length* in cm. Two research questions are of interest:

- How are *Age* and *Length* related to each other?
- What is the difference between rural/urban children w.r.t. average *Length* and does it depend on *Age*?

Now, we start with the first research question.

1. Perform a regression analysis of *Length* on *Age*.
   a. Plot *Length* against *Age* and draw the regression line in the plot. Explain what you see.
   b. What is the regression model in the sample *Length* = ...+...*Age*+*e*? (fill in the '...', i.e. the value of the regression coefficients).
   c. Complete the equation $\widehat{Length} = ...+...Age$, what does it mean, and what is the difference with b)? (fill in the '...', i.e., the value of the regression coefficients).
   d. What is the value of the intercept in the sample and what does it mean (interpret the results)?
   e. What is the value of the slope in the sample and what does it mean (interpret the results)?
   f. Determine (by means of R^2) and interpret the quality of the regression line?
   g. Plot the predicted values against the observed *Age*. What are these predicted values?
   h. Calculate by hand the predicted *Length* for a 130-month-old child and a 140-month-old child. How much do they differ in predicted length (approximately on average)?
   i. Why doesn't it make sense to determine the predicted length (based on the model) of five years old children?
2. To study the second research question, split the file in urban and rural schools.

    a. Determine a table of descriptives of *Age* and *Length*. What can you say about the observed average *Age* and *Length* of rural and urban children?

    b. Determine a scatterplot of *Length* versus *Age* for the two school types separately. Determine also the regression lines in the plots and comment on what you see.

    c. Perform the regression analysis of *Length* on *Age* for the two school types separately. Plot the predicted values of *Length* against the observed values of *Age* of the children.
Calculate the difference in predicted Length between rural and urban children of 130-months and of 140-months-old?

    d. Does it make sense to compare rural and urban children of 110-months-old (motivate your answer)? What can you say about the difference in *Length* between rural and urban children when they get older?

2.8.3 Assignment

Consider the following SPSS system file: 'Score.sav'.
  A cross-sectional study has been performed to analyse the relation between the score on a test (variable *Score*) and the time needed to study the topic (*Study_t*; in days).

    a. Calculate the correlation between *Score* and *Study_t*. Would you conclude that it is better not to study longer than say 4 or 5 days, because otherwise you may expect a lower score?

    b. Determine a plot of *Score* against *Study_t*. It appears that there were two groups. One group with much prior knowledge and another with little prior knowledge. Split the file into these two sub groups and determine the correlation again between *Score* and *Study_t* (for each of the sub groups).

    c. What would you then conclude concerning the relation between test result and study time?

    d. Comment on the statement 'It is a known fact that the correlation between alcohol consumption and criminal behaviour is $r = 0.8$. This implies that alcohol leads to criminal behaviour and hence alcohol consumption should be restricted by law'. Compare with the situation stated in the questions a-c?

    e. Can you make conclusions with respect to the relation between 'correlation' and 'causation'?

    f. Can you give another example that makes clear the difference between correlation and causation?

2.8.4 Assignment

Consider the following SPSS system file: 'Fish.sav'.

A randomised clinical trial has been conducted to analyse the effect of fish consumption (fish or meat) on bleeding time in minutes. There were 84 volunteers coming from three different research centres (Maastricht, Tromso and Zeist). Each person was measured twice: a baseline measurement before the fish/meat treatment (*Bt0* in minutes) and a postmeasurement 6 weeks after the intervention (*Bt6* in minutes).

A comparison is made between eating fish and eating meat (variable Fish) with respect to the change in bleeding time (*Bt6 − Bt0*). It was hypothesised (according to theory or what we want to find out) that the average change in bleeding time of fish eaters differs from that of the meat eaters.

The question is whether we can confirm this (research hypothesis).

  a. Produce a graph of the difference in post-pre measurement bleeding time
     (*Bt6 − Bt0*) against fish and meat eaters and describe what you see.
  b. Formulate the null – and alternative hypothesis that is being tested.
  c. Perform the appropriate test for this hypothesis and discuss your conclusion.
  d. Perform a linear regression analysis of the change in bleeding time on fish/meat eaters and compare this result with that of c.
  e. Explain why a two samples t-test (with equal variances) produces the same results as a linear regression model when the X variable is dichotomous.
  f. Which of the two techniques would you prefer in this case?

2.8.5 Assignment

Consider the following SPSS system file: 'Pulse.sav'

A preexperiment has been conducted in which a group of volunteers was asked to run at their place until they felt exhausted. Before and after the running experiment, the pulse rate (in beats/min) of each subject was measured. The difference in pulse rate is denoted by the variable *Pulse*.

It was hypothesised that the difference in pulse rate depends on the body weight (variable weight in pounds).

Analyse the data to confirm/disconfirm the above statement. Pay attention to the following.

  a. Produce a graph of *Pulse* against *Weight* and describe what you see.
  b. Formulate the null – and alternative hypothesis that is being tested.
  c. Perform a linear regression analysis of *Pulse* on *Weight* and discuss the results.

d. Discuss the quality of the regression line as a summary of the scatter plot in a.

2.8.6 Assignment

The SPSS system file 'Nutrition_missing.sav' contains data from the nutrition study, where some children do not have any measurements on *Length*.

a. Perform complete-case analysis for the linear regression of *Length* on *Age*
b. Compare the results with assignment 2.8.2 part 1.
c. Are there any differences? Motivate your answer.

2.8.7 Assignment

Consider a study with two variables X and Y, where the main interest lies in the linear association between these two variables (i.e. corr (X, Y)). Suppose about 40% of data have simultaneous missing data on both X and Y (i.e. if X is missing, Y is missing too). Which method(s) do you suggest for handling the missing data issue? Motivate your answer.

2.8.8 Assignment

Consider the following pairs of observations where some values of X are missing but Y is fully observed (missing observations are denoted by 'NA').

number	1	2	3	4	5	6	7	8	9
X	10	14	NA	15	12	16	NA	NA	18
Y	8	11	9	18	10	16	20	13	20

a. Calculate Var(X) and Corr (X, Y) using complete case analysis.
b. Impute missing values in X using the marginal mean substitution method and calculate the same quantities as part a.
c. Compare the results of part a and b.
d. Explain why Var(X) and Corr (X, Y) are lower when missing observations in X are replaced with the marginal mean substitution method (Hint: use formula 2.1 and 2.2)

2.8.9 Assignment

Consider the following SPSS system file: 'Nutrition.sav'

1. For children who are older than 130 months, remove their corresponding age values (this makes the missing data mechanism MNAR for age).

   a. Draw the scatter plot of the incomplete data with the estimated regression equation.

   b. Should this regression equation be identical to the regression equation presented in the left panel of Figure 2.18 (black line)? Discuss your answer.

   c. What do you conclude in this case?

2. For children who are taller than 140 cm, remove their corresponding *Age* values (this makes the missing data mechanism MAR for *Age*).

   a. Draw the scatter plot of the incomplete data with the estimated regression equation.

   b. Should this regression equation be identical to the regression equation presented in the right panel of Figure 2.18 (black line)? Discuss your answer.

   c. What do you conclude in this case?

2.8.10 Assignment

Consider the Nutrition study (SPSS system file: Nutrition.sav).

   a. Determine (joint and separate for rural and urban children) the observed range, average value, and standard deviation of the variables *Age* and *Length*. Comment on what you see.

   b. Produce a scatterplot of *Length* against *Age* for both rural and urban children. Also calculate the corresponding correlations for rural/urban children separately. Comment on what you see.

   c. Analyse the difference in relation between *Length* and *Age* for the two school areas. Use the following model.

   $$Length = \beta_0 + \beta_1 School_i + \beta_2 Age_i + \beta_3 School_i \times Age_i + \varepsilon_i$$

   - Determine the regression for each school area.
   - Do the relation between *Length* and *Age* statistically differ between the
   - two school areas? Discuss your conclusion.

2.8.11 Assignment

Consider the Violent-behaviour study (SPSS system file: Alc_violent.sav). See section 'short description of research and simulation study' and Section 2.3.1 for a description of this study.

Analyse the relation between alcohol consumption and violent behaviour. Is there a difference between the relationship at the group (without distinction of which subject) level and at the subject level (i.e., for each subject)? Consult the following sub-questions:

  a. Produce a scatterplot (at the group level) of *Violent* versus *Alcohol*. Describe what you see.
  b. Perform a fixed-effects linear regression analysis of *Violent* on *Alcohol* and save the unstandardised predicted values. Plot the line of predicted values versus alcohol to visualised your findings and interpret your results.
  c. Produce a scatterplot of *Violent* versus *Alcohol* for each subject. Describe what you see.
  d. Select only subject 1 and subject 3.
     - Perform a fixed-effects linear regression analysis of the relation between alcohol vs. violent behaviour at subject level (also save your unstandardised predicted values).
     - Plot the line of predicted values versus *Alcohol* per subject.
     - Compare your results with that of b. Can you explain the difference?
     - Do the relation between *Violent* and *Alcohol* differ between the two subjects?
     - Which final model would you choose?
     - Can you generalise your findings to a larger group of subjects?

2.8.12 Assignment

Consider the Growth data (SPSS system file: Growthdata.sav). (Potthoff & Roy, 1964)
  Study design:

  – Orthodontic growth measurements for 11 girls and 16 boys
  – For each subject the distance (in mm) from the centre of the pituitary gland to the pterygomaxillary fissure was recorded at ages 8,10,12,14. These two locations can be easily identified on x-ray.

Variables:

  – *Distance in mm*
  – *Sex*: 0=boy; 1=girl
  – *Age*: Age in years

Design:

Design:

Analyse the data by comparing (using the linear regression method) growth and growth velocity between boys and girls. Consult the following sub-questions:

a. Plot the <u>subject-specific profiles</u> (a plot of individual changes over time). Plot the <u>mean profiles</u> (a plot of mean changes over time, separately for boys and girls). Explain what you see. Which regression model would you specify based on these observations?

   - Perform a fixed-effects linear regression analysis of the following model

$$Distance = \beta_0 + \beta_1 Age_i + \beta_2 Sex_i + \beta_3 Age_i \times Sex_i + \varepsilon_i$$

b. Save the unstandardised predicted values.

   - Plot the line of predicted values versus age for boys and girls separately in one plot.
   - What is the interpretation of $\beta_0 + \beta_1 Age_i + \beta_2 Sex_i + \beta_3 Age_i \times Sex_i$?
   - Show that the regression parameter β_3 can be interpreted as the difference between the regression slopes of both sexes.
   - Which model describes the observed data best and argue whether there is a difference between boys and girls w.r.t. growth velocity of the *Distance*?

c. Criticise this way of analysing the data (by means of OLS method assuming uncorrelated error terms)? Recall why in the 'Violent study' the two subjects are included into the model.

2.8.13 Assignment

Consider the Violent behaviour study (SPSS system file: Alc_violent.sav)

a. Recall study design of Assignment 2.8.11. Perform a fixed-effects linear regression analysis of the relation Alcohol vs. Violent behaviour at the subject level (consider all five subjects); Include the interaction terms. Save the predicted values.

b. Plot the predicted values against *Alcohol* for each subject in one plot and discuss whether a model without interaction term will suffice.

c. Perform a fixed-effects linear regression analysis as in (a), but now without interaction terms. Which model (a or c) would you choose?

d. Argue the assumption of independent observations. Is it allowed to perform this OLS - regression analysis?

2.8.14 Assignment

Consider the Interpersonal proximity study (SPSS system file: Teacher.sav).

a. Plot the <u>subject-specific profiles</u> (a plot of proximity scores over time for each subject, separately for male and female teachers in one plot).

Plot the <u>mean profiles</u> (a plot of mean proximity over time, separately for male and female teachers). Explain what you see. Which regression model would you specify based on these observations?

b. Perform a fixed-effects linear regression analysis.

c. To find out whether there are significant differences in proximity scores over time for male and female teachers separately, first

- Split the file in males and females.

- Perform an OLS and ask for marginal means.

- Discuss whether there are significant differences.

d. Argue the assumption of independent observations. Is it actually allowed to perform the OLS – regression analysis?

How can we deal with the problem of correlated observations?

2.8.15 Assignment

Why is it generally more recommendable to use MI rather than stochastic imputation?

2.8.16 Assignment

Consider the following SPSS system file: 'Nutrition_missing.sav'
The data contains *Length, Age* and school type of children in the nutrition study. For some children, the value of *Length* is missing.

a. Impute missing values using the mean substitution, conditional mean substitution (conditional on school type and age), and stochastic regression imputation.

   b. Estimate the *Age* effect and its corresponding standard error from a linear regression of *Length* on *Age* and school type for each completed dataset in part a.

   c. Compare the results in part b. Which method has the smallest standard error? Why?

2.8.17 Assignment

In a study, the estimated fraction of missing information is 0.25. How many imputations are needed if we accept 5% loss in efficiency?

2.8.18 Assignment

Consider the SPSS system file: "Nutrition_missing.sav"

   In the nutrition study, investigate whether *Age* is the predictor of missingness in *Length*? Does the school type predict the missingness in *Length* too?

2.8.19 Assignment

Consider the SPSS system file: 'Nutrition.sav'.

   In the nutrition study, randomly remove about 20% of observations on *Age*. Suppose the analysis model is defined as the linear regression of *Length* on *Age* and school type.

   a. Write down the imputation model for *Age*.

   b. Perform an appropriate multiple imputation procedure in SPSS (or R) with $M = 10$.

   c. Compare the results of MI with that of Assignment 2.8.6, parts a and b. Are there any differences?

3

An Introduction to the Analysis of Longitudinal Data

3.1 Examples of Multilevel Designs

Before providing some examples of multilevel designs, it is important to distinguish between two sources of variation in statistical analysis methods for longitudinal data

→ *The first source of variation is due to differences in the repeated measurements of the responses for each subject over time. This type of variation is denoted as within-subject variation. The second source of variation is due to differences of the responses between the subjects at each time point. This type of variation is denoted as between-subjects variation.*

As already explained conceptually in Section 2.6.1, the between-subject variation induces correlation between repeated measurements. This type of correlation is called intraclass correlation (ICC) and will be formalised later in Section 3.3, when introducing the random-intercept model.

→ *Another type of correlation between repeated measurements may arise if each subject is measured by an observer (possibly self-reporting) and each observed measurement is correlated with the previous observed measurement(s) conditional on each subject. This type of correlation is known as serial correlation, which is different from the ICC.*

This version of correlation will also be elaborated later in this chapter.

Multilevel designs can be broken down into two parts: a subclass of cross-sectional multilevel designs and a class of repeated measures including longitudinal designs, where subjects are measured repeatedly over time. An example of a cross-sectional multilevel design is a design where pupils come from randomly selected classes within randomly selected schools that are evaluated on their performance. An example of a repeated measures design is a study where subjects are repeatedly measured under several drug conditions in order to compare these conditions with respect to some health outcomes.

In this book, focus will be on the longitudinal designs, although the techniques that we discuss can be applied to any form of multilevel designs.

3.1.1 Cross-Sectional Multilevel Designs

3.1.1.1 Performance of Pupils in Schools

A typical example used to explain multilevel structures is a cluster randomised study on the performance of pupils in two different educational programs. A random sample of 20 schools, assigned to a new educational program based on a guided-learning system, is compared with a random sample of 25 schools with a conventional education program. Note that the number of schools of the two groups may differ. Within each school, a random sample of parallel classes (different sizes allowed) was collected. The performance on an exam of all pupils within each class was evaluated.

This is an example of a three-level design, where pupils are nested within randomly sampled classes, and classes are nested within randomly sampled schools. The fact that classes and schools are considered to be randomly sampled makes it possible to generalise classes and schools to the population from which these classes and schools are sampled. The first-level observations are the performances of pupils. The second-level observations are classes, and the third-level observations are schools. To prevent misunderstandings, the different observed classes and schools of the variables *Class* and *School, respectively,* will be denoted as categories of these variables.

> → *The fact that the classes and schools are randomly sampled, makes the variables Class and School random. The nesting structure of the random variables Class and School (Class nested within Schools) determines the number of levels.*

Suppose that the performance of an arbitrary chosen pupil does not affect the performance of any other pupil, so that all pupils are assumed to do the exam completely on their own. It then can be argued that the performance of two arbitrarily chosen pupils within each class and/or school is more alike than that of pupils from different classes or schools. The reason for this is that different classes may have different teachers with different teaching skills. Consequently, there may be differences in performance between different classes. Different schools may also differ from each other due to, for example, different educational policies. Consequently, due to different performances in classes and/or schools, the performance of pupils within each class will be correlated, that is correlation will exist if we consider all (unconditionally) classes and schools. We say that there is correlation between pupils, which is induced by the difference between classes and/or schools. The same also applies to the classes. The classes within each school are also correlated due to the difference between the schools. The correlation may become more pronounced, as the difference between classes (schools) increases.

It should be noted that conditional on each class, the performance of pupils is uncorrelated, that is if we restrict to exactly one class. Furthermore, it is also plausible to assume that, conditional on each school, the performance of the classes is uncorrelated. The cases at the highest level (schools in this example) must be assumed to be independent.

3.1.1.2 Cross-Country Comparison: Cross-Sectional Design

Another example concerns the effect of a large-scale anti-alcohol campaign. A random sample of seven anti-alcohol-campaign countries is compared with eight other non-anti-alcohol countries. Within each country, a random sample of approximately 3000 respondents is sampled. Furthermore, the countries are subdivided into northern and southern countries. The dependent variable expressed in terms of change (difference) in alcohol consumption after and before the campaign was measured for each respondent.

This is an example of a two-level design. Three thousand respondents are nested within randomly sampled countries. This makes it possible to generalise the results to other countries than those already included in the sample. Note that the countries are independently sampled and that conditional on each country, the respondents are independently sampled. The first-level observations are respondents and the second-level observations are countries.

Although the countries are nested within the intervention status, the variable *Alcohol Campaign* (with values anti and non-anti) is not considered as a level. The variable *Geographic Position* (with values north and south) is also not a level, although countries are nested within geographic position. It is therefore important to recognise whether a variable can be considered random or not. If, for example, northern and southern area (geographic position) could be considered as a random draw from all possible positions (e.g. north, north east, south east, ... etc.), *Geographic Position* can be considered as a level (and thus, it would be a third-level design).

3.1.2 Longitudinal Designs

3.1.2.1 Cross-Country Comparison: Longitudinal Design

In the above anti-alcohol campaign study, suppose that alcohol consumption per subject is measured three times; each year once within a study period of three years. We then have a three-level longitudinal design with *Time* as the first-level observation, *Subject* as the second level and *Country* as the third level.

In general, the number of levels in a multilevel design is determined by the number of random factors for which the categories are a random sample from all possible factor categories. For example, a sample may be drawn from all possible countries, subjects and schools. If the observed levels are not considered as a random sample, the factor is fixed.

The subjects in the present study are a random sample from some well-defined population of subjects (within each country). The results of an analysis, based on the subjects in the sample, will then be generalised to the population. *Subject* will be treated as a random factor. However, in some (rare) situations, interest is only to draw conclusions with respect to the subjects in the sample, which makes the results of the statistical analysis not possible to be generalised to a larger group and the *Subject* as a factor will be considered fixed.

The reverse situation is also possible. Suppose that one is only interested in the results about these 15 countries and generalisation to other countries is not relevant. Suppose further that there are (non-negligible) differences between countries. The usual way to handle this type of heterogeneity is to include *Country* as a discrete variable in the regression model with *Treatment* by *Country* interactions. The idea is that inferences can be made for each country. However, this intention to include *Country* as a covariate does not make sense because by conditioning on *Country* one automatically conditions on *Treatment* due to the cluster randomisation. By specifying *Country* as a level, this problem can be solved.

Variables such as *Geographic Position, Treatment Status, Social Economic Status, Sex* and *Marital Status* are usually considered fixed factors, because interest is only restricted to the observed categories of such variables.

If there is only one country in the sample, all subjects can be assumed to be uncorrelated. In the present study, however, the subjects are nested within the presumed randomly sampled countries. Alcohol behaviour of the subjects within countries is more alike than that of respondents between countries. For example, the amount of alcohol consumption in the Netherlands is, on average, much lower than in France. Consequently, given an observation of a French person, it is possible to predict better than change whether an observation of another person is French or Dutch. This induces for an arbitrary country correlation between observations at the (second) subject level. We will see later that the amount of correlation is determined by the amount of between-country variability. The same reasoning can also be given regarding the repeated measurements within each subject (compared with Figure 2.13). The differences between countries and subjects induce correlation between first-level observations.

→ *Observations at the lower levels tend to be correlated due to the multilevel structure.*

3.1.2.2 Growth Study

The next example is a study of orthodontic growth of 11 girls and 16 boys, which was first analysed by Potthoff and Roy (1964). In this study, the distance (in mm) from the centre of the pituitary gland to the pterygomaxillary fissure was recorded at four different ages: 8, 10, 12 and 14. This is an example of a two-level longitudinal design. The first-level observations are the repeated measurements nested within subjects, that is each subject is measured several times (four times). The second-level observations are the subjects that are considered to be independently and randomly sampled. Since measurements within subjects are more alike than measurements between subjects, there is a correlation induced by the multilevel structure.

3.1.2.3 Interpersonal Proximity Study

The last example is a study concerning the evaluation of teacher's interpersonal behaviour in the classroom. We gave a short description of this study in Section 2.4.1. The Proximity study is also an example of a two-level longitudinal design with missing observations. The first level consists of the four time points (occasions) that are repeatedly measured for each subject, that is the variable *Occasion* nested within *Teacher*. If we want to generalise the teachers to a larger group (population), *Teacher* must be considered random and therefore is a second-level factor.

Another second-level variable is *Sex* (0 = male teacher; 1 = female teacher). *Sex* could possibly be a predictor of the proximity score of the teacher. It is also possible that *Sex* moderates the relationship between the measurement occasion and the proximity score (i.e. an interaction between *Sex* and *Occasion* may be considered).

In this study, it is likely that respondents remember their proximity scores of the preceding years, leading to serial correlations between repeated measurements in addition to the correlation induced by the multilevel structure. Note that serial correlation due to memory effects or other factors that cause a time lag can also be present in the previous two examples. In this book, we will approach the issue of serial correlation exclusively for the Proximity study. More examples of longitudinal designs will be presented in Chapter 4 and further.

The next section compares a standard linear regression analysis, assuming independent observations with a multilevel regression analysis that takes into account correlated observations. A longitudinal data set is used for this comparison.

3.2 Comparison of the Multilevel Linear Regression Model with the Standard (OLS) Linear Regression

Two important characteristics in a two-level longitudinal design should be kept in mind:

1. Decomposition of variation between subjects and within subjects. In Figure 3.1, two regression lines for the regression of *Distance* on *Age* are presented, separately for two subjects. For each subject, the small double arrows indicate the deviation of observations from the corresponding subject-specific regression line. The variance of these deviations is called the within-subject variance. The large double arrows indicate the deviation of the subject-specific regression lines from the average regression line (the line in the middle). The corresponding variance is called the between-subjects variance.

→ *In longitudinal designs (in contrast to cross-sectional multilevel designs), the between-subjects variation is often larger than the within-subject' variation.*

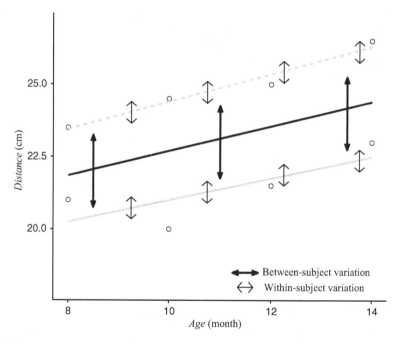

FIGURE 3.1
An illustration of two sources of variation in longitudinal designs.

2. The observations at different time points are correlated. An incorrect way to deal with correlated data is to neglect the multilevel structure and serial correlation. Consequently, all observations are treated as if they were uncorrelated. In that case, the standard OLS method would then be applicable. We will discuss by means of an example the consequences of applying the standard OLS method to a multilevel structure with respect to point estimates and corresponding standard errors of the regression parameters.

The Growth study, which is described in Section 3.2.1, serves as an example for the comparison between a multilevel and an OLS model. Note that the design is balanced in time (i.e. all children were measured at the same time points) and there are no missing observations. We also assume that no serial correlation exists, which implies that the repeated observations are uncorrelated conditional on each child. Figure 3.2 shows the scatter plot and individual profiles of the children for this study. From this figure, it can be argued that the between-subjects variation is considerably larger than the within-subject variation and that there is (induced) correlation between distance measurements over time points.

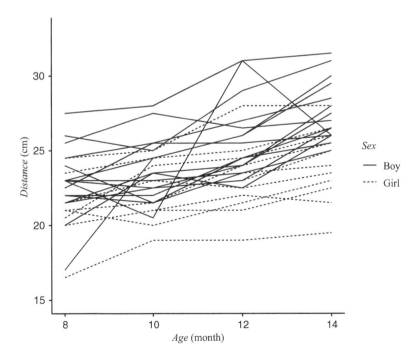

FIGURE 3.2
Individual profiles in the Growth study identifying sex of child.

Despite the dependency of the repeated measurements, suppose that the OLS regression analysis is (incorrectly) performed. Figure 3.3 shows the mean profiles for boys and girls. Suppose that we would like to compare boys and girls with respect to their growth of the *Distance* and that we are only interested in the linear trend over time.

Moreover, suppose that we would like to test whether there is an interaction between *Sex* and *Age*. The corresponding fixed-effects multiple linear regression model is then:

$$Distance_{ij} = \beta_0 + \beta_1\, Sex_i + \beta_2\, Age_{ij} + \beta_3\, Sex_i \times Age_{ij} + V_{ij}, \qquad (3.1)$$

with β_0 as intercept and β_p, $p = 1, 2, 3$ are the regression parameters and V_{ij} is defined as a new notation of the error term for child i at time point (age) j, $i=1,\ldots, n$, $j=1,\ldots, m$. It is assumed that the error term is normally distributed with mean 0 and constant variance σ_v^2. (compared with model 2.22) with the index 'ij' for child i at time point (age) j. This notation is also used in standard statistical packages such as SPSS (IBM, 2020) and SAS (SAS Institute Inc., 2016). Note that σ_v^2 indicates the deviations of the observations around the sex-specific regression lines. For the OLS analysis, all V_{ij}s are assumed to be

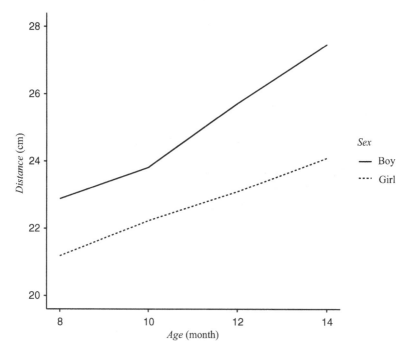

FIGURE 3.3
Average profiles for each sex group in the Growth study.

TABLE 3.1

Growth Study: Comparison between the OLS and Correct Analysis

Parameter	Estimate	Std. Error	df	*t*-value	*p*-value	95% Confidence Interval	
						Lower Bound	Upper Bound
a) OLS							
Intercept	16.34	1.42	104	11.54	0.000	13.53	19.15
Age	0.78	0.13	104	6.22	0.000	0.53	1.03
Sex	1.03	2.22	104	0.47	0.643	−3.37	5.43
Sex × Age	−0.30	0.20	104	−1.54	0.126	−0.70	0.09
b) Correct analysis (random-intercept model)							
Intercept	16.34	0.98	103.99	16.65	0.000	14.39	18.29
Age	0.78	0.08	79.00	10.12	0.000	0.63	0.94
Sex	1.03	1.54	103.99	0.67	0.504	−2.02	4.08
Sex × Age	−0.30	0.12	79.00	−2.51	0.014	−0.55	−0.06

Dependent Variable: *Distance*.

independently and identically distributed. In Equation 3.1, *Sex* is coded as boy = 0 and girl = 1, *Age* in years and *Distance* in cm.

Table 3.1 shows the results of the OLS analysis (part a) where the data are treated as if they were uncorrelated (i.e. neglecting the multilevel structure). The results of the correct analysis that considers the correlation in the data is reported in part b of Table 3.1. The correct analysis, the random-intercept model, acknowledges the multilevel structure of the data so that no independence assumption is made. As we will see in Section 3.3.3, the random-intercept model is a better choice to describe the growth data than the standard OLS method.

Comparing both results in Table 3.1, the following observations can be made: First, the point estimates of the OLS method are identical to those of the appropriate method. Pleasurable as it may seem, this situation only occurs when all subjects are measured at the same time points (balanced in time) and when there are no missing observations. The parameter estimates of the OLS method will be biased for unbalanced (in time) design and/or when missing observations occur. Second, the standard errors of the estimates from both methods are not the same. To explain the discrepancy, let us take a closer look at the sex-specific age effect and the age-specific sex effect.

3.2.1 Age-Specific Sex Effect

An age-specific sex effect reflects the difference in average *Distance* between boys and girls for each age group separately. In terms of the regression parameters, this difference can be calculated by substituting the codes for the variable *Sex* into the estimated model of Equation (3.1) and then subtracting both averages as follows.

$$\widehat{Distance}_{female} = \hat{\beta}_0 + \hat{\beta}_1 \times 1 + \hat{\beta}_2 \times Age + \hat{\beta}_3 \times 1 \times Age$$
$$\widehat{Distance}_{male} = \hat{\beta}_0 + \hat{\beta}_1 \times 0 + \hat{\beta}_2 \times Age + \hat{\beta}_3 \times 0 \times Age,$$

(3.2)

The age-specific sex effect is the difference

$$\widehat{Distance}_{female} - \widehat{Distance}_{male} = \hat{\beta}_1 + \hat{\beta}_3 \times Age$$

The estimated value $\widehat{Distance}$ for males and females is approximately equal to the average distance for large samples.

Table 3.2 a shows the estimates of age-specific sex effect with the corresponding standard errors. The OLS estimates of the standard errors are biased downward. This can be understood as follows. An additional correlated (repeated) measurement has less information than an additional uncorrelated measurement. For example, if two repeated measurements are highly correlated (perhaps perfectly correlated), there will be hardly any (or no) additional information in the second measurement on top of what is already available from the first measurement. In contrast, if the two measurements are uncorrelated, there will be more additional information from the second measurement. Consequently, if we add many measurements and incorrectly assume independency, then it looks as if the total amount of information will increase. Since standard errors reflect the amount of information and become smaller when more information is available, the standard errors from OLS estimation on repeated measurements become smaller than they actually should be. For the degrees of freedom, the opposite happens. Assuming that two correlated measurements are uncorrelated, they both are free to vary,

TABLE 3.2

Growth Study: Comparison between OLS and an Appropriate Model

		Appropriate Analysis (Random Intercept)	OLS
a) Estimated age-specific sex effects (with standard errors)			
Age = 8	sex effect	−1.41 (0.84)	−1.41 (0.74)
Age = 10	sex effect	−2.02 (0.77)	−2.02 (0.48)
Age = 12	sex effect	−2.63 (0.77)	−2.63 (0.48)
Age = 14	sex effect	−3.24 (0.84)	−3.24 (0.74)
b) Estimated sex-specific age effects (with standard errors)			
Sex = 0	age effect	0.78 (0.08)	0.78 (0.13)
Sex = 1	age effect	0.48 (0.09)	0.48 (0.15)

while actually only one is free to vary, because the other is bounded by the correlation between both measurements. This is why larger degrees of freedom are found for OLS estimation of correlated measurements.

→ *Therefore, by treating correlated observations (over time) as independent observations, the standard error of the estimated sex effect for each age level will be too low and the degrees of freedom will be too high.*

Consequently, the corresponding p-values will become too small leading to an increase of type I error rates.

3.2.2 Sex-Specific Age Effect

A sex-specific age effect reflects the amount of growth for boys and girls separately. In terms of regression parameters, this effect can be determined by substituting the codes of the variables into the regression Equation (3.1) leading to

$$\widehat{Distance}_{t+1} - \widehat{Distance}_t = \begin{cases} \hat{\beta}_2 + \hat{\beta}_3 \text{ if } \text{Sex} = 1 \\ \hat{\beta}_2 \quad\;\; \text{if } \text{Sex} = 0 \end{cases} \tag{3.3}$$

The corresponding standard errors are shown in Table 3.2b. The OLS estimates of the standard errors are biased upward. This can be understood as follows. Using OLS, the standard error of the estimated age effect is calculated based on the combined between- and within-subject variability. For the estimation of the amount of change over time, however, one should only account for the within-subject variability. Differences between subjects should not affect the precision of this estimate.

→ *Hence, the sex-specific standard errors of the estimated age effects will be too large when using OLS.*

Consequently, the corresponding p-values will be too large leading to a loss of power.

The interaction term in Equation (3.1) can be interpreted as the difference in growth velocity between boys and girls and should be estimated accounting for only the within-subject variability. This situation is comparable with that of the sex-specific age effect.

→ *Consequently, the standard error of the interaction term and the corresponding p-values will be over-estimated when using OLS.*

In Table 3.1, it is shown that the OLS method leads to a nonsignificant contribution of the interaction term (p-value = 0.126), whereas the appropriate analysis method leads to a highly significant contribution (p-value = 0.014). Consequently, the analysis of the Growth study with the incorrect OLS method leads to the wrong conclusion that the growth velocity of the boys is not significantly different to that of the girls.

3.3 Accounting for the Multilevel Structure

3.3.1 Formulation of the Problem: A Marginal and Subject-Specific Representation

Recapitulating the problem of analysing longitudinal data, Figure 3.4 illustrates two ways to model longitudinal data:

1. Marginal model, which can be interpreted as a regression line through means
2. Subject-specific model, which can be interpreted as the mean of subject-specific regression lines.

We emphasise that both models result in identical average regression line (i.e., both bold lines are identical)

We start with the marginal model and define, for the sake of argument and simplicity, the regression model of *Distance* on *Age* by

$$Distance_{ij} = \beta_0 + \beta_1 Age_{ij} + V_{ij}. \tag{3.4}$$

Because of the two-level structure of the growth data, the V_{ij}'s are correlated. In particular, each of the six pairs of four repeated measurements and the variances of the four repeated measurements must be estimated. However, the measurements between different children are assumed to be uncorrelated, which is crucial to ensure the unbiasedness of the estimates of the regression parameters. In Equation 3.4, in addition to the regression parameters β_0 and β_1, 10 other parameters for the variances and correlations of the error terms must be estimated: four variances (for Age = 8, 10, 12 and 14) and six covariances (and thus the correlations; see Equation 2.3), that is, the covariances between Age = 8 and Age = 10, between Age = 8 and Age = 12 and up to between Age = 12 and Age = 14. One way to account for this is to estimate all model parameters using Restricted Maximum Likelihood estimation (REML), which is available in many statistical packages (see the accompanying website for a short manual on longitudinal data analysis with SPSS and R). Unfortunately, estimating all variance-covariance parameters

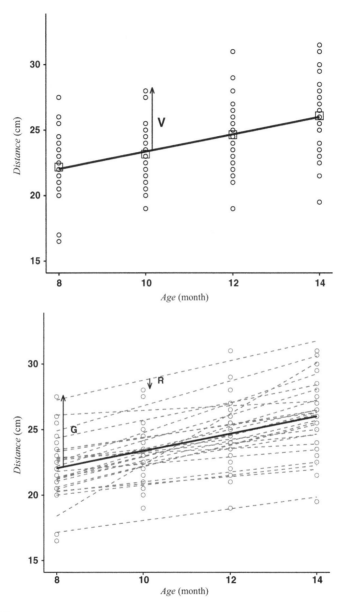

FIGURE 3.4
Regression line of means (above) and mean of regression lines (below).

(we denote this as 'unstructured') may lead to unstable results (Verbeke & Molenberghs, 2000). Our experience shows that unstable results may occur if the number of time points is larger than three. With three time points, we need to estimate six variance-covariance parameters. Estimating all parameters will not lead to unstable parameter estimates, but the situation with four or more time points may be more problematic. In Chapter 4, we offer some basic guidelines about the best choice of the variance-covariance structure, that is simpler variance-covariance structures with less parameters and without a significant loss of information. A procedure is proposed that compares several candidates of variance-covariance structures using objective criteria. In this book, we will follow the basic guidelines as explained in Chapter 4 to find the most parsimonious model without a significant loss of information, except if there are three repeated measurements. The variances and covariances will then remain unstructured.

> → *In conclusion, an approach to analyse data from a multilevel design using Equation 3.4, is known as a 'Marginal approach'. The regression line will be fitted through the graph of average Distance as a function of Age. Details of the marginal model will be discussed in Section 3.4.*

A second option is to opt for a so-called 'subject-specific approach' (see the second plot of Figure 3.4). The idea of using the subject-specific approach consists of two stages. First, a regression line is specified for each subject, and second, the average of all subject-specific regression lines is obtained.

> → *Both approaches serve the same purpose, i.e., to estimate the same regression parameters unbiasedly. It can be shown that the regression line through the means (first plot) is the same as the mean of all subject-specific regression lines (second plot).*

3.3.2 Fixed or Random Factor

Suppose that we repeat the data sampling of, for example the Growth study in another comparable study with the same research problem and the same methodology with respect to the sampling procedure. Is it likely, on the one hand, that the same set of subjects will be sampled or collected? If the answer is no, then the variable *Subject* should be considered a random factor. If, on the other hand, the same set of subjects will be collected, *Subject* should be considered fixed.

What can we say about *Sex*? It is likely that when repeating the data collection, we will still end up with the same two categories of boys and girls (but with different boys and girls) and no other gender types (at least what was relevant for this Growth study). Hence, *Sex* is considered to be a fixed factor. Since we want to compare growth between boys and girls, *Sex* should be included into the model as an independent variable. In this case, *Sex* is a second-level independent variable (because it only sub-divides the second-level subjects into boys and girls). Without going into detail, treating *Subject* as a fixed independent variable, it is still possible to estimate the interaction *Sex × Age* (difference in speed of growth between boys and girls) unbiasedly. However, as discussed in Chapter 2, treating *Subject* as a fixed independent variable is not a good strategy, as it has major drawbacks. For example, in the Growth study, no generalisation can be made to a larger group of children. Moreover, *Subject* should not be treated as fixed if there are serial correlations because the independence assumption in the OLS method is violated (observations of each subject are correlated). It will also be hard to evaluate the *Sex* effect, because basically the model is fitted for each child and thus also *Sex* separately. Finally, this model cannot deal with time-dependent covariates.

What we should do when a variable (like the variable *Subject*) is considered to be a random factor is the topic of the next section.

3.3.3 Treating Subjects as a Random Factor

Usually, the purpose of a study is to generalise the findings to a larger group (population) of subjects than those in the sample. This problem can be solved by assuming that the subjects in the study form a random sample from a certain known population and adding this information into the specification of the regression model. The variable *Subject* is then treated as a random factor. In addition, specifying subject's differences comes down to indicating in what respect the subjects differ from each other. The simplest model in which subjects can differ from each other is with respect to their intercept only. Because *Subject* is considered to be a random factor, the intercept is also considered to be random. This type of model is called a random-intercept model. If the subjects also differ from each other regarding their regression slopes in addition to the intercepts, the model is called a random-slope model. We will elaborate both models subsequently.

3.3.3.1 Random-Intercept Models

The random-intercept model allows us to model subject-specific intercepts. In most statistical packages, these intercepts are assumed to be realisations from a normal distribution. Consider, for the sake of argument, the following simplified model for the growth data.

$$Distance_{ij} = \beta_{0i} + \beta_1 Age_{ij} + R_{ij},$$

where $\beta_{0i} = \beta_0 + G_{0i},$

and

$R_{ij} \sim N(0, \sigma^2)$ as the error term $\left(\begin{array}{l}\text{with } \sigma^2 \text{ as the first level}\\ \text{within} - \text{subject variance}\end{array}\right)$ (3.5)

$G_{0i} \sim N(0, \tau_0^2)$ as the random intercept $\left(\begin{array}{l}\text{with } \tau_0^2 \text{ as the second level}\\ \text{between} - \text{subject variance}\end{array}\right)$

G_{0i} and R_{ij} are uncorrelated

Equation 3.5 is visualised in Figure 3.5. For the ease of presentation, we do not distinguish between boys and girls at this moment. In this example, we assume that the subject-specific regression lines only differ in intercept. That is, why the lines in Figure 3.5 are parallel with slope equal to β_1, which can be interpreted as the average growth velocity of the children. Furthermore, the regression intercepts are supposed to be a random sample

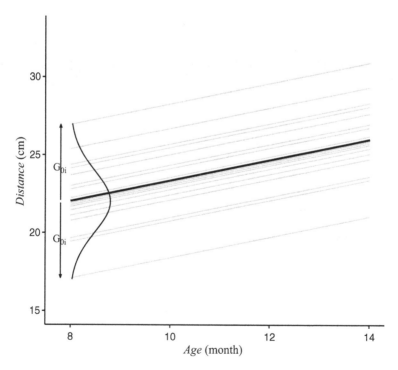

FIGURE 3.5
A graphical representation of the random-intercept model.

from a normal distribution. The different intercepts are specified in Equation 3.5 as β_{0i}, $i = 1,\dots,n$, which is denoted as random intercept, normally distributed with a mean value β_0 and variance τ_0^2.

> → *The first-level (within-subject) variance σ^2 indicates for each subject how far the observations are from its corresponding regression line, which is assumed to be equal for all subjects. The second-level (between-subject) variance τ_0^2 indicates how far the subject-specific regression lines are from the average regression line.*

The random-intercept model specifies all subject-specific regression lines. All individual regression lines are connected to each other through the underlying assumption of the random intercept and the functional relationship as specified in the model. The REML method (e.g. Diggle et al., 2013) is used to estimate all model parameters. The average profile of interest is determined using all available information contained in the data.

Usually, inferences are only made about the average regression line. Inferences about the individual regression lines are often not of main interest and it will not be discussed in this book. Equation 3.5 can be rewritten in a form that is similar to the standard (fixed) regression specification by decomposing the responses into a fixed part of averages and a random part as follows,

$$
\begin{aligned}
Distance_{ij} &= \beta_{0i} + \beta_1\, Age_{ij} + R_{ij}, \\
Distance_{ij} &= \beta_0 + \beta_1\, Age_{ij} + \left(R_{ij} + G_{0i} \right) \\
Distance_{ij} &= \beta_0 + \beta_1\, Age_{ij} + V_{ij}, \\
&\text{where } V_{ij} = R_{ij} + G_{0i} \\
&\text{with} \\
&Var\left(V_{ij} \right) = \sigma^2 + \tau_0^2 \\
corr\left(V_{ij}, V_{ik} \right) &= ICC = \frac{\tau_0^2}{\sigma^2 + \tau_0^2},
\end{aligned}
\tag{3.6}
$$

The term ICC is the intraclass correlation. Just like in the fixed-effects situation, the response $Distance_{ij}$ can be decomposed into the systematic fixed part $\beta_0 + \beta_1 Age$ (average regression line) and the random part V_{ij} with overall variance σ_v^2, which is a marginal model representation of the random-intercept model. The overall variance σ_v^2 is a combination of the first-level within-subjects variance (σ^2) and the second-level between-subjects variance(τ_0^2).

> → *In fact, any random-effects model can be expressed as a marginal model. The other way around, however, is not always possible.*

In general, the specification of the systematic part is completely similar to the fixed regression case. However, the error terms V_{ij} are correlated. In Equation 3.6, it can be shown that the error terms are correlated with a constant correlation equal to ICC $= \tau_0^2 / (\sigma^2 + \tau_0^2)$ (see e.g., Snijders and Bosker, 2012).

→ *ICC reflects the proportion of the total amount of variance explained by the differences between subjects.*

If the between-subject variance τ_0^2 is very large relative to the within subject variance (σ^2), the numerator of ICC (τ_0^2) will be approximately equal to the denominator ($\sigma^2 + \tau_0^2$). The value of the ICC will then be close to one implying that the error terms between time points are highly correlated.

Usually, these variances and covariances (or correlations) are summarised in a table that consists of m rows and m columns (called an $m \times m$ matrix). Each of the m rows or columns represent a different time point. The element in the j-th row and k-th column is equal to cov (V_{ij}, V_{ik}) or $corr$ (V_{ij}, V_{ik}). In particular, the correlation between the responses at the same time point j is equal to 1 and the covariances are equal to the between-subject variance τ_0^2.

→ *in the random-intercept model, the correlations between time points are equal for all subjects, so that we can drop the subscript i.*

In the literature, such matrix is denoted as a covariance (or correlation) matrix $\mathbf{V}_{cov}$ (or $\mathbf{V}_{corr}$). If the number of time points is equal to $m = 4$, the matrix $\mathbf{V}_{corr}$ can be written as

$$\mathbf{V}_{corr} = \begin{pmatrix} 1 & \dfrac{\tau_0^2}{\sigma^2 + \tau_0^2} & \dfrac{\tau_0^2}{\sigma^2 + \tau_0^2} & \dfrac{\tau_0^2}{\sigma^2 + \tau_0^2} \\[3mm] \dfrac{\tau^2}{\sigma^2 + \tau_0^2} & 1 & \dfrac{\tau_0^2}{\sigma^2 + \tau_0^2} & \dfrac{\tau_0^2}{\sigma^2 + \tau_0^2} \\[3mm] \dfrac{\tau_0^2}{\sigma^2 + \tau_0^2} & \dfrac{\tau_0^2}{\sigma^2 + \tau_0^2} & 1 & \dfrac{\tau_0^2}{\sigma^2 + \tau_0^2} \\[3mm] \dfrac{\tau_0^2}{\sigma^2 + \tau_0^2} & \dfrac{\tau_0^2}{\sigma^2 + \tau_0^2} & \dfrac{\tau_0^2}{\sigma^2 + \tau_0^2} & 1 \end{pmatrix}, \tag{3.7}$$

or in terms of the corresponding matrix of variances and covariances

$$\mathbf{V}_{corr} = \begin{pmatrix} \sigma^2 + \tau_0^2 & \tau_0^2 & \tau_0^2 & \tau_0^2 \\ \tau_0^2 & \sigma^2 + \tau_0^2 & \tau_0^2 & \tau_0^2 \\ \tau_0^2 & \tau_0^2 & \sigma^2 + \tau_0^2 & \tau_0^2 \\ \tau_0^2 & \tau_0^2 & \tau_0^2 & \sigma^2 + \tau_0^2 \end{pmatrix}. \tag{3.8}$$

This type of a structure of a correlation (covariance) matrix is called compound symmetry or exchangeable. The covariances between time points are equal and fully determined by the between-subject variance τ_0^2 (i.e. induced by the two-level design).

In general, $\mathbf{V}_{cov}$ is a matrix that consists of overall variances and covariances of the responses.

> → *The variances and covariances of the random effects are summarised in a so-called $\mathbf{G}_{cov}$ matrix, whereas those of the measurement errors are summarised in a so-called $\mathbf{R}_{cov}$ matrix. The overall variances and covariances (the elements of $\mathbf{V}_{cov}$) are combinations of the variances and covariances of the (level two) random effects of the within-subject (level one) measurement errors.*

Table 3.3 summarises the notations of various variance – and covariance – components for a random-intercept (RI) model. In this RI model, the subjects only differ in their intercepts. Hence, the level-two $\mathbf{G}_{cov}$ matrix only consists of the random intercept variance $\mathbf{G}_{cov} = (\tau_0^2)$ (See Table 3.3, third column). The level-one observations are supposed to be uncorrelated with equal variances over time (i.e., no serial correlations are imposed). The corresponding $\mathbf{R}_{cov}$ matrix for the random-intercept model has covariances equal to zero and constant variance equal to σ^2.

TABLE 3.3

Variance-Covariance Components of the Random-Intercept Model with $m = 4$ Time Points

Variances and Covariances	Matrix	RI
Level one (within-subjects): Measurement error and no serial correlations	$\mathbf{R}_{cov}$	$\begin{pmatrix} \sigma^2 & 0 & 0 & 0 \\ 0 & \sigma^2 & 0 & 0 \\ 0 & 0 & \sigma^2 & 0 \\ 0 & 0 & 0 & \sigma^2 \end{pmatrix}$
Level two (between-subjects): Random effects variances and covariances	$\mathbf{G}_{cov}$	$\begin{bmatrix} \tau_0^2 \end{bmatrix}$
overall	$\mathbf{V}_{cov}$	$\begin{pmatrix} \sigma^2+\tau_0^2 & \tau_0^2 & \tau_0^2 & \tau_0^2 \\ \tau_0^2 & \sigma^2+\tau_0^2 & \tau_0^2 & \tau_0^2 \\ \tau_0^2 & \tau_0^2 & \sigma^2+\tau_0^2 & \tau_0^2 \\ \tau_0^2 & \tau_0^2 & \tau_0^2 & \sigma^2+\tau_0^2 \end{pmatrix}$

→ *In a random-intercept model without serial correlation, all elements of the* $\mathbf{R}_{cov}$ *matrix should be added by* τ_0^2 *to obtain the* $\mathbf{V}_{cov}$ *matrix (see equation (3.8)).*

The correctness of the random-intercept model depends partly on the plausibility of the restrictions based on substantive arguments. If we know, for example, that the *Distance* was measured by a well-calibrated electronic measurement equipment, then serial correlation is unlikely to exist.

Let us consider the full growth model as given in Equation (3.1). The REML method was performed to obtain unbiased estimates of all model parameters. A random-intercept model seems plausible, because the subject-specific profiles (see Figure 3.4 second plot) were more or less parallel with considerable between-subjects variability. Part of the output is depicted in Table 3.4. Note that the covariance matrices are presented here in a table format.

The random-intercept variance is estimated as $\hat{\tau}_0^2 = 3.30$, which is the only element of the $\mathbf{G}_{cov}$ matrix. The $\mathbf{R}_{cov}$ matrix has equal residual variances estimated by $\hat{\sigma}^2 = 1.92$. Note that all covariances are equal to zero, because no serial correlations were assumed. The corresponding overall correlation matrix $\mathbf{V}_{corr}$ is equal to

$$\mathbf{V}_{corr} = \begin{pmatrix} 1 & 0.63 & 0.63 & 0.63 \\ 0.63 & 1 & 0.63 & 0.63 \\ 0.63 & 0.63 & 1 & 0.63 \\ 0.63 & 0.63 & 0.63 & 1 \end{pmatrix} \tag{3.9}$$

The estimated correlation $Corr(\hat{V}_{ij}, \hat{V}_{ik})$ between time points j and k is $\hat{\tau}_0^2 / (\hat{\sigma} + \hat{\tau}_0^2) = 0.63$. Apparently, the observations are fairly highly correlated

TABLE 3.4

Growth Study: The $\mathbf{G}_{cov}$ - and $\mathbf{R}_{cov}$- Matrix of the Random Intercept Model

$\mathbf{G}_{cov}$ matrix

	$\hat{\beta}_{0i}$
$\hat{\beta}_{0i}$	3.30

$\mathbf{R}_{cov}$ matrix

	[*Age* = 8]	[*Age* = 10]	[*Age* = 12]	[*Age* = 14]
[*Age* = 8]	1.92	0	0	0
[*Age* = 10]	0	1.92	0	0
[*Age* = 12]	0	0	1.92	0
[*Age* = 14]	0	0	0	1.92

($r = 0.63$) solely due to the two-level structure. The within-subjects residual variance ($\hat{\sigma}^2 = 1.92$) is smaller than the between-subject variance ($\hat{\tau}_0^2 = 3.30$) as expected from Figure 3.4. The estimates of the fixed parameters are shown in Table 3.1 (the correct analysis).

In summary, it can be concluded that:

- The elements of $\mathbf{R}_{cov}$ are the level-one within-subject variances and covariances indicating the discrepancy between the observations and the subject-specific regression lines (see Figures 3.4 and Figure 3.5). The same amount of discrepancy is assumed for all subjects (Verbeke and Molenberghs, 2000). In Figure 3.6, a schematic representation of the random-intercept model is given separately for boys and girls. The $\mathbf{R}_{cov}$ matrix has equal diagonal elements ($\hat{\sigma}^2 = 1.92$, see Figure 3.6) and zero off diagonals.

- The elements of $\mathbf{G}_{cov}$ are the level-two between-subject variances and covariances indicating the discrepancy between the subject-specific regression lines and the average regression line. In Figure 3.6, $\mathbf{G}_{cov}$ of the growth model also consists of one element ($\hat{\tau}_0^2 = 3.30$), which is the same for the boys and for the girls.

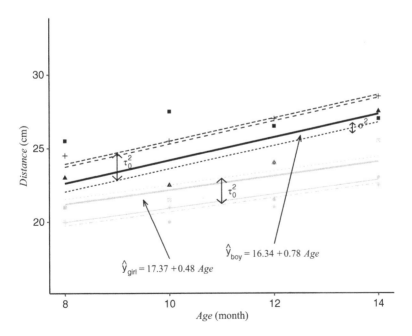

FIGURE 3.6
Growth study: Graphical representation of the random-intercept model for boys and girls separately

- The elements of $\mathbf{V}_{cov}$ are the overall variances and covariances of the responses indicating the discrepancy between the observations and the average regression line. For the random-intercept model, the $\mathbf{V}_{cov}$ is given in Table 3.3 (last row).

3.3.3.2 Random-Slope Models

In many cases, individual regression lines may reveal different regression slopes. In that situation, there will be a so-called 'Time × Subject' interaction, where *Subject* is a random factor. To model different regression slopes, a random slope can be added to the random-intercept model. A random-effects model with a random intercept and slope is:

$$
\begin{aligned}
Distance_{ij} &= \beta_{0i} + \beta_{1i} Age_{ij} + R_{ij}, \\
\text{where} \quad \beta_{0i} &= \beta_0 + G_{0i}, \text{and} \\
\beta_{1i} &= \beta_1 + G_{1i}
\end{aligned}
$$

$R_{0i} \sim N\left(0,\sigma^2\right)$ as the error term

$G_{0i} \sim N\left(0,\tau_0^2\right)$ as the random intercept

$G_{1i} \sim N\left(0,\tau_1^2\right)$ as the random slope

τ_{01} as the covariance between random intercept and random slope

G_{0i} and G_{0i} are uncorreated with R_{ij}

(3.10)

Equation 3.10 indicates that the subject-specific regression lines differ in intercept and slope. The subject-specific intercepts and slopes are supposed to be samples from normal distributions. The random intercept is specified in Equation (3.10) as β_{0i}, $i = 1,...,n$ with a mean value β_0 and variance τ_0^2. The different slopes are indicated by β_{1i}, $i = 1,...,n$, which is denoted as a random slope, normally distributed with a mean value β_1 and variance τ_1^2. Moreover, the random intercept (RI) and the random slope (RS) are supposed to be correlated with covariance equal to τ_{01}. Table 3.5 summarises both $\mathbf{R}_{cov}$ and $\mathbf{G}_{cov}$ matrices for a random-effects model with m = 4 time points when no serial correlations are assumed (column 3). Following Snijders and Bosker (2012, p. 76.), the overall variances and covariances of the responses (the $\mathbf{V}_{cov}$ matrix) are also determined by:

$$
\begin{aligned}
var\left(Y_{it}\right) &= \tau_0^2 + 2\tau_{01}t + \tau_1^2 t^2 + \sigma^2 \\
cov\left(Y_{ij}, Y_{ik}\right) &= \tau_0^2 + \tau_{01}\left(j+k\right) + \tau_1^2\left(j \times k\right)
\end{aligned}
$$

(3.11)

with j, k = 1,...,m, m = total number of time points. No serial correlations are assumed in this example. All covariances between the responses (see Equation (3.11)) are due to the multilevel design.

TABLE 3.5

Variances- Covariances of the Random-Intercept/Slope Model and AR(1) for $m = 4$ Time Points

Variances and Covariances	Matrix	RI/RS	AR (1) Heterogeneous For Balanced Data
Level one: Measurement error and serial correlations	$\mathbf{R}_{cov}$	$\begin{pmatrix} \sigma^2 & 0 & 0 & 0 \\ 0 & \sigma^2 & 0 & 0 \\ 0 & 0 & \sigma^2 & 0 \\ 0 & 0 & 0 & \sigma^2 \end{pmatrix}$	$\begin{pmatrix} \sigma_1^2 & \rho\sigma_1\sigma_2 & \rho^2\sigma_1\sigma_3 & \rho^3\sigma_1\sigma_4 \\ \rho\sigma_1\sigma_2 & \sigma_2^2 & \rho\sigma_2\sigma_3 & \rho^2\sigma_2\sigma_4 \\ \rho^2\sigma_1\sigma_3 & \rho\sigma_2\sigma_3 & \sigma_3^2 & \rho\sigma_3\sigma_4 \\ \rho^3\sigma_1\sigma_4 & \rho^2\sigma_2\sigma_4 & \rho\sigma_3\sigma_4 & \sigma_4^2 \end{pmatrix}$
Level two: Random effects variances and covariances	$\mathbf{G}_{cov}$	$\begin{pmatrix} \tau_0^2 & \tau_{01} \\ \tau_{01} & \tau_1^2 \end{pmatrix}$	Unspecified
overall	$\mathbf{V}_{cov}$	a 4 x 4 matrix with diagonal and off-diagonal elements given in equation (3.11)	$\mathbf{V}_{cov} = \mathbf{R}_{cov}$

The formulation of random-effects models usually assumes that no serial correlations exist. This type of correlation is not due to the multilevel structure, but to the fact that the error of a measurement on a subject at a particular point in time somehow correlates with the error of measurement at another (possibly adjacent) point in time. This may be due to memory effect or other reasons that cause a carryover in time within a subject.

3.3.3.3 Serial Correlation

Differences between subject's characteristics, such as subject-specific intercepts and/or slopes, induce correlation between repeated measurements. The difference in intercept, for example, may cause correlation that is proportional to the number of differences between subjects. In the random-effects model mentioned previously, it is assumed that the $\mathbf{R}_{cov}$ matrix contains zero covariances conditional on the subjects.

The most common serial correlation structure in practice is the AR (1) (first-order auto regressive) structure. It assumes that the amount of correlation between two time points j and k only depends on the length of the interval between these time points, that is the correlation $R_{jk} = \rho \mid j - k \mid$, with $-1 < \rho < 1$ and $j, k = 1, \ldots, m$. Consequently, the AR (1) serial correlations will be closer to zero, as the interval length between two time points increases.

Note that the $\mathbf{R}_{cov}$ matrix with serial correlations has off-diagonal elements unequal to zero.

3.3.3.4 Example: Random-Effect Model or Marginal Model With and Without Serial Correlation

Consider the proximity study as an example how to use the random intercept/ slope model with and without serial correlation. The description of this study is already given in Section 2.4.1.

Following the discussion of Section 2.4.2, we specify the regression Equation 3.12 for this study. We emphasise that the systematic part of the model treats occasion as a categorical variable because of the nonlinear relationship between *Proximity* and *Time*.

We are mainly interested in the comparison of average proximity scores between the different observed occasions in this study. Because an interaction could be expected between *Occasion* and *Sex*, the following regression model (marginal representation) is specified (cf. Section 2.4.2).

$$
\begin{aligned}
Proximity_{ij} = {} & \beta_0 + \beta_1 Occ_{0ij} + \beta_2 Occ_{1ij} + \beta_3 Occ_{2ij} + \beta_4 Sex_i \\
& + \beta_5 Occ_{0ij} \times Sex_i + \beta_6 Occ_{1ij} \times Sex_i + \beta_7 Occ_{2ij} \\
& \times Sex_i + V_{ij},
\end{aligned}
\tag{3.12}
$$

The dependent variable $Proximity_{ij}$ is the proximity score of subject i at occasion $j = 0,\ldots,3$. The systematic part consists of:

- the variables Occ_{kij}, $k = 0, 1, 2$, which are dummy variables for occasion j;
- the variable Sex_i, which is coded 0 for males and 1 for females;
- and the interaction terms $Occ_{kij} \times Sex_i$, $k = 0, 1, 2$ representing the *Sex* by *Occasion* interaction.

The random terms V_{ij} indicate the deviations of the observed values around the average regression line per teacher's *Sex*. It is also assumed that the variances attributed to these deviations are the same for male teachers and female teachers. In principle, however, this assumption can be relaxed if necessary.

The structure of the overall variances and covariances ($\mathbf{V}_{cov}$ matrix) should be established before estimating the regression parameters, because the estimated standard errors of the regression parameters could be biased if an incorrect covariance structure is imposed. Different variance-covariance structures can be specified. Before trying any available variance-covariance structure, we could try to make an educated guess by first looking at the scatterplot of the observed profiles of all subjects as shown in Figure 3.7. Looking at the scatterplot, it can be seen that there is a considerable amount of variability between subjects, suggesting the presence of a second-level intercept variance τ_0^2. Moreover, the amount of variability seems to decrease with time, suggesting that a random-intercept model is not sufficient to describe the data properly. A model with a random intercept and a random slope might therefore be a better choice with the second-level

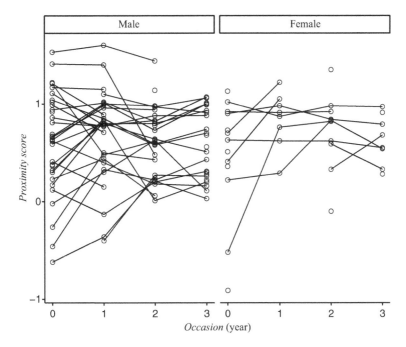

FIGURE 3.7
Individual profiles for male and female teachers in the Proximity study.

slope variance τ_1^2. Covariation between the random intercept and random slope is also possible.

On the other hand, serial correlations are expected because the students evaluate the performances of the teachers on a yearly basis, and the scores at preceding occasions might influence the scores on the proximity scale in the following occasions, leading to correlated errors. By means of an example, two competing covariance structures will be modelled:

1. A marginal regression model with AR (1) serial correlations with heterogeneous variances.
2. A subject-specific regression model with a random intercept and a random slope and with no serial correlations.

Tables 3.6a and 3.6b show a part of the output of an AR (1) and a random-intercept/slope model without serial correlation, respectively. In Table 3.6a, the variances and covariances of the responses ($\mathbf{V}_{cov}$ matrix) are presented. In SPSS, for example, this output can be obtained by specifying the $\mathbf{R}_{cov}$ matrix (within-subject variances and covariances) to be AR (1) heterogeneous and leaving the random components ($\mathbf{G}_{cov}$ matrix) unspecified. As a result, the $\mathbf{V}_{cov}$ matrix equals the $\mathbf{R}_{cov}$ matrix (see Table 3.5, last column).

TABLE 3.6A

Proximity Study: V_{cov} Matrix of Equation 3.12 with AR (1) Serial Correlations and Heterogeneous Variances

	A Table Representation of the V_{cov} Matrix			
	Occ_0	Occ_1	Occ_2	Occ_3
Occ_0	0.32	0.20	0.11	0.08
Occ_1	0.20	0.22	0.12	0.09
Occ_2	0.11	0.12	0.12	0.09
Occ_3	0.08	0.09	0.09	0.13

TABLE 3.6B

Proximity Study: G_{cov} Matrix and R_{cov} Matrix of the Random-intercept/Slope Model (no serial correlations)

G_{cov} Matrix		
	$\hat{\beta}_{0i}$	$\hat{\beta}_{8i}$
$\hat{\beta}_{0i}$	0.24	−0.05
$\hat{\beta}_{8i}$	−0.05	0.02

R_{cov} matrix				
	Occ_0	Occ_1	Occ_2	Occ_3
Occ_0	0.05	0	0	0
Occ_1	0	0.05	0	0
Occ_2	0	0	0.05	0
Occ_3	0	0	0	0.05

Looking at the elements of the V_{cov} matrix in Table 3.6a, it can be seen that the variances decrease with time, which is also observed from the plot in Figure 3.7. Moreover, the covariances will move toward zero, as the interval width between the two time points increases. This is in line with an AR (1) structure.

Note that in model specification 3.12, the variable *Occasion* is treated as a discrete variable.

> → *To specify a random slope, we assume that the deviation of the subject-specific slope around the line through the averages is changing linearly (Verbeke & Molenberghs, 2000).*

This linearly changing deviation can be accomplished by adding a continuous variable Z with the same values as the discrete variable *Occasion* with values 0, 1, 2, 3 and with random parameter β_{8i}, normally distributed with mean zero and variance τ_1^2, indicating the subject-specific slope. If the continuous variable Z were not added, then each category of the variable *Occasion* will be treated as a separate variable.

The corresponding **G** matrix will then consist of 15 parameters to be estimated (10 for the variances and covariances between the dummy variables, four for the covariances between the residual and the four dummy variables and one residual variance), while the **V** – matrix, which consists of a combination of the **G**- and the **R**- matrix, only consists of 10 degrees of freedom. Consequently, the model will not be identified (this can be checked by performing the analysis and compare with the G matrix from Table 3.6b, where a continuous variable Z is added instead).

With a continuous variable Z, the reader may check that the V_{ij} in Equation 3.12 will be $V_{ij} = G_{0i} + G_{1i} Z_{ij} + R_{ij}$.

When a random-intercept/slope model is analysed, the $\mathbf{V}_{cov}$ matrix can be calculated from the $\mathbf{G}_{cov}$ -and the $\mathbf{R}_{cov}$ -matrix by using the formula in Equation (3.11). It turns out that the $\mathbf{V}_{cov}$ and $\mathbf{V}_{corr}$ matrices are estimated as

$$\mathbf{V}_{cov} = \begin{pmatrix} 0.28 & 0.18 & 0.13 & 0.08 \\ 0.18 & 0.20 & 0.11 & 0.08 \\ 0.13 & 0.11 & 0.14 & 0.08 \\ 0.08 & 0.08 & 0.08 & 0.12 \end{pmatrix}$$

$$\mathbf{V}_{corr} = \begin{pmatrix} 1 & 0.78 & 0.66 & 0.44 \\ 0.78 & 1 & 0.68 & 0.51 \\ 0.66 & 0.68 & 1 & 0.58 \\ 0.44 & 0.51 & 0.58 & 1 \end{pmatrix} \tag{3.13}$$

From Equation (3.13), it can be seen that the variances of the responses decrease with time, and the covariances (and correlations) also decrease as the interval width between two time points increases. It should be noted that the form of the $\mathbf{V}_{cov}$ matrix in Equation (3.13), which shows decreasing variances and covariances, is not typical for a random-intercept/slope model. If hypothetically speaking, the same study was performed in a different period than between occasion 0–3 (keeping all estimated regression parameters constant), say starting between occasion 1 and occasion 4, then the $\mathbf{V}_{cov}$ matrix could be different. Consequently, the scatterplot of the dependent variable against time can have different shapes depending on which period the study is performed. Figure 3.8 shows the three different possibilities:

a. The observations spread as time passes.
b. The observations close together as time passes.
c. The observations cross as time passes.

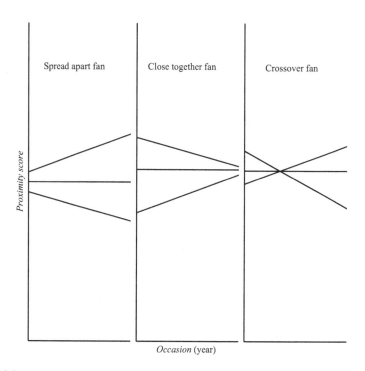

FIGURE 3.8
Three different fan shapes of the random-slope model.

There are no other possibilities due to the quadratic nature and the positive values of the variances of the random effects. For the sake of comparison, suppose that this Proximity study was performed between occasion 1 and occasion 4. The $\mathbf{V}_{cov}$ components would still show decreasing covariances, but the variances are not decreasing and resemble a crossover fan situation as shown in Figure 3.8c. The corresponding $\mathbf{V}_{cov}$ and $\mathbf{V}_{corr}$ matrices would be

$$
\mathbf{V}_{cov} = \begin{pmatrix} 0.20 & 0.11 & 0.08 & 0.04 \\ 0.11 & 0.14 & 0.08 & 0.06 \\ 0.08 & 0.08 & 0.12 & 0.07 \\ 0.04 & 0.06 & 0.07 & 0.13 \end{pmatrix}
$$
$$
\mathbf{V}_{corr} = \begin{pmatrix} 1 & 0.68 & 0.51 & 0.27 \\ 0.68 & 1 & 0.58 & 0.42 \\ 0.51 & 0.58 & 1 & 0.56 \\ 0.27 & 0.42 & 0.56 & 1 \end{pmatrix}
$$

(3.14)

In Equation (3.14), the variances first decrease and increase again at the last occasion.

If the study was performed between occasion 3 and occasion 6, the variances (Equation 3.15) would increase in agreement with a spreading fan as mentioned in Figure 3.8a. Note that in this case, the covariances do not decrease.

$$
\mathbf{V}_{cov} = \begin{pmatrix} 0.12 & 0.07 & 0.07 & 0.06 \\ 0.07 & 0.13 & 0.10 & 0.11 \\ 0.07 & 0.10 & 0.17 & 0.15 \\ 0.06 & 0.11 & 0.15 & 0.25 \end{pmatrix}
$$

$$
\mathbf{V}_{corr} = \begin{pmatrix} 1 & 0.56 & 0.46 & 0.37 \\ 0.56 & 1 & 0.63 & 0.60 \\ 0.46 & 0.63 & 1 & 0.74 \\ 0.37 & 0.60 & 0.74 & 1 \end{pmatrix}
$$

(3.15)

In all situations, the correlation of the residuals decreases as the distance between the time points increases. Let us focus on the results of the analysis of the observed occasions (0, 1, 2, 3), because that is the situation as it is observed. For both the AR (1) heterogeneous model and the random-intercept/random slope model, all estimates of the variances and the covariances are significant at 0.05 significance level. This can be seen from the Wald test depicted in the Tables 3.6a and 3.6b. Both models are competing alternatives to describe the variance-covariance structure of the responses. These tests, however, do not provide any evidence favouring one model over another. Some selection criteria on how to choose the most suitable model will be discussed in Chapter 4. If serial correlation AR (1) is added in top of the random intercept/random slope model, the reader may check that the parameter estimates do not change much and the $\mathbf{V}_{cov}$ remains more or less the same.

3.4 Marginal Models

The model with Equation 3.4 can be estimated properly (efficiently and without bias) if the structure of the matrix $\mathbf{V}$ is known. In practice, however, $\mathbf{V}$ will be unknown and needs to be estimated. As discussed earlier, one way to accomplish this is to use the subject-specific approach, which specifies the within-subject variances and between-subject variances and the correlations through the $\mathbf{R}$ - and $\mathbf{G}$- matrices, respectively. Individual plots such as in Figure 3.7 can help us to formulate the competing $\mathbf{R}$ –and $\mathbf{G}$ - matrices. These competing matrices can then be compared statistically, which will be described in detail in Chapter 4. Once the best choice has been selected, the corresponding overall $\mathbf{V}$ matrix can be determined that has the form of Equation (3.7) (or 3.8) for the random-intercept model and 3.11 for the random-intercept/slope model.

As noted, every random-effects model with corresponding **R**-and **G**-matrices can be represented by a marginal model in the form of Equation (3.4).

In general, matrix algebra is required to determine the **V** matrix, given the **R**- and **G** –matrices. For a random-intercept or random-intercept/slope model with an additional AR (1) serial correlation, the **V** matrix can be calculated as follows:

→ *Random intercept (**G**$_{cov}$ matrix = (τ_0^2)) combined with serial correlation (**R**$_{cov}$ matrix = AR (1), homogeneous or heterogeneous): τ_0^2 should be added to each of the elements of the **R**$_{cov}$ matrix.*

→ *Random intercept/slope (**G**$_{cov}$ matrix = $\begin{pmatrix} \tau_0^2 & \tau_{01} \\ \tau_{01} & \tau_1^2 \end{pmatrix}$) combined with serial correlation (**R**$_{cov}$ matrix = AR(1), homogeneous): add each diagonal element of the **R**$_{cov}$ matrix representing the variance at time point t to $var(Y_{it}) = \tau_0^2 + 2\tau_{01}t + \tau_1^2 t^2 + \sigma^2$, and add each off-diagonal element of **Rcov** matrix representing the covariances at time point j and k to $cov(Y_{ij}, Y_{ik}) = \tau_0^2 + \tau_{01}(j+k) + \tau_1^2(j \times k)$.*

There are more overall variance-covariances than those induced by random effects and/or AR (1). Most statistical packages (like SAS, R, and SPSS) provide several correlation/covariance structures for **V**. Thus, instead of following a subject-specific, one can also fit a marginal model as specified by Equation (3.4) using one of the pre-specified **V** matrices, or create a user-specified **V** matrix (like in SAS).

There are several alternatives. Some well-known structures that are used in this book are (examples with four time points):

- *Independent (Scaled Identity or DIAG):*
  *there is no correlation between measurements. In fact, with **V** = Scaled Identity, Equation 3.4 reduces to the standard (OLS) model as described in Chapter 2. If **V** = DIAG, the regression parameters can also be obtained using a weighted least square (Draper & Smith, 1966, p. 108). There are two possibilities. Homogeneous with equal variances (scaled Identity) and Heterogeneous with unequal variances (DIAG).*

$$\mathbf{V}_{corr} = \begin{pmatrix} 1 & 0 & 0 & 0 \\ 0 & 1 & 0 & 0 \\ 0 & 0 & 1 & 0 \\ 0 & 0 & 0 & 1 \end{pmatrix} \tag{3.16}$$

- *Compound symmetry (CS or CSH):*
  the correlation between each pair of measurements is assumed constant. There are two possibilities. Homogeneous with equal variances (CS) and Heterogeneous with unequal variances (CSH).

$$\mathbf{V}_{\text{corr}} = \begin{pmatrix} 1 & \rho & \rho & \rho \\ \rho & 1 & \rho & \rho \\ \rho & \rho & 1 & \rho \\ \rho & \rho & \rho & 1 \end{pmatrix} \tag{3.17}$$

- *Unstructured (UN):*
  the correlation between each pair of measurements is assumed to be different. The variances may also be different.

$$\mathbf{V}_{\text{corr}} = \begin{pmatrix} 1 & \rho_1 & \rho_2 & \rho_3 \\ \rho_1 & 1 & \rho_4 & \rho_5 \\ \rho_2 & \rho_4 & 1 & \rho_6 \\ \rho_3 & \rho_5 & \rho_6 & 1 \end{pmatrix} \tag{3.18}$$

- *First-order autoregressive (AR (1) or ARH (1)): autoregressive structure of time-decaying correlation of equal spaced repeated measurements. There are two possibilities. Homogeneous with equal variances (AR (1)) and Heterogeneous with unequal variances (ARH (1)).*

$$\mathbf{V}_{\text{corr}} = \begin{pmatrix} 1 & \rho & \rho^2 & \rho^3 \\ \rho & 1 & \rho & \rho^2 \\ \rho^2 & \rho & 1 & \rho \\ \rho^3 & \rho^2 & \rho & 1 \end{pmatrix} \tag{3.19}$$

- *Spatial power (sp-power) (generalisation of AR (1) for unequal spaced repeated measurements):*

$$\mathbf{V}_{\text{corr}} = \begin{pmatrix} 1 & \rho^{(t_2-t_1)} & \rho^{(t_3-t_1)} & \rho^{(t_4-t_1)} \\ \rho^{(t_2-t_1)} & 1 & \rho^{(t_3-t_2)} & \rho^{(t_4-t_2)} \\ \rho^{(t_3-t_1)} & \rho^{(t_3-t_2)} & 1 & \rho^{(t_4-t_3)} \\ \rho^{(t_4-t_1)} & \rho^{(t_4-t_2)} & \rho^{(t_4-t_3)} & 1 \end{pmatrix} \tag{3.20}$$

- *Toeplitz (TP or TPH):*
  the correlation between measurements is assumed to change with time span for equal spaced repeated measurements. There are two possibilities: homogeneous with equal *variances (TP) and Heterogeneous with unequal variances (TPH).*

$$
\mathbf{V}_{corr} = \begin{pmatrix} 1 & \rho_1 & \rho_2 & \rho_3 \\ \rho_1 & 1 & \rho_1 & \rho_2 \\ \rho_2 & \rho_1 & 1 & \rho_1 \\ \rho_3 & \rho_2 & \rho_1 & 1 \end{pmatrix} \tag{3.21}
$$

- *First-order ante dependence (AD (1)):*
  A correlation structure that can be used for unequal spaced measurements. The correlations between adjacent time points are unrestricted. However, the other correlations are determined by the correlation between adjacent time points.

$$
\mathbf{V}_{corr} = \begin{pmatrix} 1 & \rho_1 & \rho_1\rho_2 & \rho_1\rho_2\rho_3 \\ \rho_1 & 1 & \rho_2 & \rho_2\rho_3 \\ \rho_1\rho_2 & \rho_2 & 1 & \rho_3 \\ \rho_1\rho_2\rho_3 & \rho_2\rho_3 & \rho_3 & 1 \end{pmatrix} \tag{3.22}
$$

Let us consider the analysis of the growth data. Instead of a random-intercept model, one can also perform a marginal model with a compound symmetry **V** matrix. Table 3.7 shows some results.

TABLE 3.7

Growth Study: A Marginal Model with a Compound Symmetry **V** – matrix

						95% Confidence Interval	
Parameter	Estimate	Std. Error	df	t-value	p-value	Lower Bound	Upper Bound
a)							
Intercept	16.34	0.98	103.99	16.65	0.000	14.39	18.29
Sex	1.03	1.54	103.99	0.67	0.504	−2.02	4.08
Age	0.78	0.08	79.00	10.12	0.000	0.63	0.94
Sex × Age	−0.30	0.12	79.00	−2.51	0.014	−0.55	−0.06

Dependent Variable: *Distance.*

V Matrix	[Age = 8]	[Age = 10]	[Age = 12]	[Age = 14]
b)				
[Age = 8]	5.22	3.30	3.30	3.30
[Age = 10]	3.30	5.22	3.30	3.30
[Age = 12]	3.30	3.30	5.22	3.30
[Age = 14]	3.30	3.30	3.30	5.22

Compound Symmetry.

Comparing this marginal model with the random-intercept model, the following can be said. The parameter estimates as displayed in Table 3.7a are exactly the same as those based on the random-intercept model (see Table 3.1, second table). Furthermore, the estimate of the overall variance (Table 3.7b) and the covariances (correlations) of the compound symmetry model are equal to 5.22 and 3.30 (0.63), respectively, which is also exactly equal to the estimated overall variance of the random-intercept model. Apparently, for the analysis of the growth data, the random-intercept model leads to exactly the same results as the compound symmetry model. In general, however, the random-intercept model does not always lead to the same results as the compound symmetry model (see assignment 3.6.4 and Section 4.2.3).

3.5 Handling Missing Observations in Longitudinal Designs

In Chapter 2, we discussed how to handle missing data in cross-sectional studies. Here, we will focus on the issue of missing data in longitudinal studies and discuss different approaches to deal with them. The first step is to obtain an overview of the missing data patterns by counting how many of them are encountered. We then compare longitudinal and cross-sectional designs with respect to the methods developed to address the issue of missing data. An important distinction must of course be made because, unlike cross-sectional designs, data are correlated in longitudinal studies so that this aspect should be taken into account when considering any missing data method.

3.5.1 Patterns of Missing Data in Longitudinal Studies

It is always useful to study patterns of missing data in longitudinal studies because it facilitates evaluation of the missing data mechanisms and appropriateness of the missing data methods. Moreover, we advise to distinguish variables playing different roles in longitudinal studies as follows:

1. Dependent variable: a variable measuring the outcome of interest across time.
2. Time-varying independent variable(s): those measured at each time point and serve as covariates or predictors in the main analysis
3. Baseline or time-invariant independent variable(s): those measured once, typically at baseline.

In the Growth study, the dependent variable of interest is the distance between the centre of the pituitary gland to the pterygomaxillary fissure, which is measured at four time points. *Age* is the time-varying independent variable showing the child's age at each time point, and finally, *Sex* is the time-invariant independent variable that is measured at baseline.

When there are no missing observations in time-invariant independent variables, the most common patterns of missing data in longitudinal studies are monotone (see Figure 1.3). This pattern implies that one or more subjects leave the study permanently at a particular time point, and therefore, the dependent variable (and possibly time-varying independent variables) is missing for them. As an example, the Alzheimer study was a clinical trial involving patients with Alzheimer's disease aiming at comparing placebo with two active treatments (Verbeke and Molenberghs, 2006, par.17.18). The dependent variable was dementia score measured at baseline and over six occasions after treatment. Baseline variables involved sex and age of the patients. In this study, the missing data patterns are monotone, as the dementia score had missing observations only, and all baseline variables were fully observed. Table 3.8 summarises the patterns for this study ('O' and 'M' represent being observed and missing, respectively).

Missing observations in the Alzheimer study cause seven distinct patterns in the data. The largest pattern includes about 66% of subjects for whom the dementia score is fully observed, that is those who completed the study. In contrast, about 4% of subjects dropped out after the first measurement of dementia score were included.

For such monotone patterns, it is less likely that subjects dropped out totally by chance. As a result, the MCAR mechanism might be excluded from the possible missing data mechanisms. Although the possibility of MNAR mechanism cannot be ruled out in general, we still may assume (or at least

TABLE 3.8

Alzheimer Study: Patterns of Missing Data from 344 Patients

Pattern group	Baseline	Week1	Week2	Week4	Week6	Week8	Week10	%
1	O	O	O	O	O	O	O	66%
2	O	O	O	O	O	O	M	3%
3	O	O	O	O	O	M	M	5%
4	O	O	O	O	M	M	M	8%
5	O	O	O	M	M	M	M	10%
6	O	O	M	M	M	M	M	4%
7	O	M	M	M	M	M	M	4%
Observed	100%	96%	92%	82%	74%	67%	66%	

start with) the missing data mechanism that is MAR in the sense that the probability of observing the dementia score at a particular week depends on the dementia scores in preceding weeks, but not on the score of the current week (and future weeks). When the patterns of missing data are monotone, and the MAR assumption is deemed plausible, standard multilevel or marginal models introduced earlier in this chapter can safely be used because such methods allow for an unbalanced design in the data (as missing observations of dementia score make the study design unbalanced).

In longitudinal studies, the non-monotone patterns of missing data can occur. Let us revisit the Proximity study where teachers' interpersonal behaviour was evaluated at four occasions. In this study, *Sex* and *Occasion* were fully observed as baseline and time-varying independent variables, respectively, but the proximity score is missing for some teachers in an arbitrary manner. Figure 3.9 depicts the patterns of missing data for this study. Here, the proximity score can be observed at a particular occasion even after the score was not observed at earlier occasions. For instance, the third pattern shows that the proximity score was observed at baseline and occasions 2 and 3 but not at occasion 1.

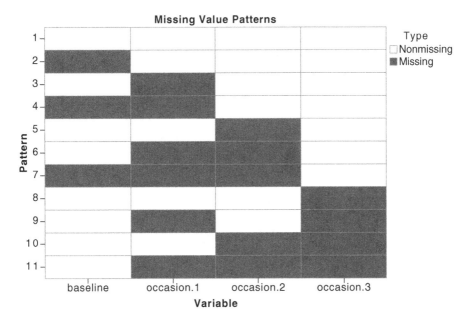

FIGURE 3.9
Patterns of missing data in the Proximity study.

→ *With non-monotone patterns of missing data, the probability of observing the dependent variable at a particular time point can depend on the dependent variable at earlier time points, but due to the intermittent pattern, those dependent variable values (at earlier time points) might be missing too. As a result, it is hard to argue the MAR assumption for non-monotone missing data patterns in general. The MNAR assumption might thus be more convincing for non-monotone patterns of missing data.*

The abovementioned statement does not mean that the MAR or even MCAR assumption is impossible with the non-monotone patterns of missing data as discussed in the studies by Robins and Gill (1997) and Little (1994).

When both the dependent and baseline variable(s) have missing observations, the patterns of missing data are often non-monotone. In such cases, a relatively straightforward method is to first impute (multiply) missing data in the baseline variables (but not the dependent variable) and then apply the standard methods for longitudinal data in the imputed datasets.

3.5.2 Complete-Case Analysis in Longitudinal Studies

Complete-case analysis (CCA) in longitudinal studies implies that only subjects with complete records contribute to the analysis. As an example, CCA uses only 66% of the subjects in the Alzheimer study because only these patients have all intended measurements (see pattern 1 in Table 3.8). As a result, the estimate of regression weights may lose efficiency and the impact on statistical power could be dramatic, but no bias is introduced in the estimates as long as the missing data mechanism is MCAR. This is because the complete cases are assumed to be a random sample of all the data. However, it is typically hard to justify the MCAR mechanism in longitudinal studies.

A less stringent version of MCAR is what is known as a covariate-dependent mechanism (Little, 1995). Here, the missing data mechanism can only be related to fully observed independent variables, which implies the missingness is independent of the dependent variable at any time point conditional on the independent variables. Therefore, covariate-dependent missingness can be thought of as an MCAR mechanism with conditional independence. This raises a subtle, but important, point as discussed in Fitzmaurice et al. (2012).

→ *Under covariate-dependent missingness, if all variables related to missingness are included in the analysis as independent variables, CCA will be valid and therefore delivers unbiased estimates, although potentially inefficient.*

This is a very important special case, because all longitudinal studies with a monotone pattern essentially have more missing data as time goes on, which makes time related to missingness. If conditional on time, missingness is unrelated to the dependent variable, inclusion of time in the analysis validates the MCAR mechanism (or the covariate-dependent mechanism), and hence CCA is valid. However, the mechanism is no longer MCAR if time is not included in the CCA as an independent variable. Hedeker and Gibbons (2006) illustrated this point by means of a small simulation study, where the missingness depended on time and time by treatment group interaction.

Several tests have been proposed in the literature to evaluate the MCAR assumption, including the covariate-dependent missingness mechanism. The Little's test (Little 1988) is perhaps the most recognised test of the MCAR mechanism. These tests, however, are often misinterpreted by data analysts and researchers because these are not simply tests of the MCAR against non-MCAR assumption. For instance, the Little's test provides evidence for the MCAR assumption relative to a MAR alternative hypothesis. Thus, nonsignificant results of the Little's test do not imply that an analysis assuming MCAR is justified! This is because the Little's test does not distinguish the MCAR from MNAR mechanism. As Rhoads (2012) notes, 'Tests of the MCAR assumption should be used only if the MAR assumption is deemed likely to hold a priori.'

→ *We therefore conclude that the observed data cannot prove the MCAR mechanism is true, that is, the MCAR assumption cannot solely be tested by the data at hand. However, the MCAR assumption can be falsified by the observed data. We also note that falsification of the MCAR assumption does not provide any justification for the MAR assumption as the observed data alone can never provide evidence for or against the MAR assumption.*

In contrast to MCAR or its covariate-dependent version, CCA delivers biased estimates if the missing data mechanism is MAR (e.g. missingness depends on the observed outcome measurements). This is because the remaining complete cases are no longer a random sample of the whole population. This is confirmed by many through theoretical arguments and simulation studies (see, for instance, Tan, Jolani and Verbeek, 2018). Similar to cross-sectional studies, the bias of CCA under MNAR (for the dependent variable) will be even more severe than under MAR because subjects with complete data are a selective group of population (i.e., not a random sample of the whole population).

Perhaps, a seemingly obvious approach to evaluate plausibility of the MCAR assumption is to perform an analysis using CCA (including variables related to missingness) and compare it with the linear mixed or marginal regression models that essentially assume MAR. If the results are different, we can falsify the MCAR assumption. However, if the results are comparable, it still cannot be concluded that the MCAR assumption is validated, because the missing data mechanism could be either MCAR or MNAR.

3.5.3 Last Observation Carried Forward in Longitudinal Studies

A heuristic method to deal with missing data in longitudinal studies is last observation carried forward (LOCF). The method is simple and accommodates both monotone and non-monotone patterns of missing data. For each subject, LOCF replaces the missing value of each time point with the last previously observed value and therefore reproduces the original rectangular matrix. Table 3.9 illustrates the method for three cases in the Proximity study. As an example, the missing proximity score for the first case at occasion 2 (denoted by '.') is replaced by the last previously observed proximity score at occasion 1.

The LOCF approach is attractive because it enables to include all individuals into the analysis, and standard complete data methods can directly be applied. The method, however, is widely criticised for various reasons (see, e.g., Wood, White and Thompson, 2004, Kenward and Molenberghs, 2009; Lane, 2008). The key concern is the stringent underlying assumption of LOCF. As noted in the study by White, Carpenter and Horton (2012), a valid LOCF analysis (in randomised studies) assumes that the subject's unobserved independent variable is constant over time (for monotone patterns) or during the unobserved period until the next observed period (for non-monotone patterns). This is nevertheless often unrealistic and hard to justify in practice. It should also be noted that this assumption has no logical connection with the MCAR assumption so that it cannot be used to validate the LOCF approach. Another consideration originates from the standpoint of imputation strategy. LOCF relies only on a single subject to fill in the missing observation(s) of that subject, while any other imputation methods (including multiple imputation) additionally borrow information from other cases with complete data. As a result, LOCF is not optimal in the sense that all available information is not used when missing values are filled in.

Sometimes, LOCF is defended for being a conservative estimate of the treatment effect in randomised studies. This is not necessarily true because the direction and magnitude of the bias are difficult to predict, as it depends on the true but unknown treatment effect (Molenberghs and Kenward, 2007;

TABLE 3.9

Proximity Study: Results of LOCF for a Subset of Subjects. The Dots are Missing and the Imputed Values Using LOCF Are in Bold

	Observed and Missing Data				Last Observation Carried Forward			
ID	Baseline	Occ1	Occ2	Occ3	Baseline	Occ1	Occ2	Occ3
1	0.41	1.05	.	0.91	0.41	1.05	**1.05**	0.91
2	0.64	.	.	.	0.64	**0.64**	**0.64**	**0.64**
3	1.13	.	1.35	.	1.13	**1.13**	1.35	**1.35**

Siddiqui, Hung and O'Neill, 2009). Other variants such as baseline observed carried forward share similar limitations and generally are not advised.

3.5.4 Advanced Missing Data Methods in Longitudinal Studies

This section studies two advanced methods, namely, direct likelihood (DL) and multiple imputation (MI) for handling missing data in longitudinal studies. The usage and appropriateness of these methods depend on the following two factors:

1. location of missing observations and
2. reason(s) for missing observations.

We discuss the properties of these state-of-the-art methods with respect to these two factors. Where needed, the methods will also be compared with CCA.

3.5.4.1 Missing Observations in the Dependent Variable Only

In longitudinal studies, it is very common to be confronted with missing observations in the dependent variable only. This is because the dependent variable of interest is often measured repeatedly over the course of the study. For example, the dementia score in the Alzheimer study has not been measured completely leading to an incomplete data matrix (i.e. all intended measurements were not collected for every subject). In such situations, the standard random-effects models introduced earlier in this chapter can be used since they allow for data from unbalanced designs. However, these models assume that the missing data mechanism is ignorable in the sense that the inference is only based on the observed data without addressing the missing data mechanism (see Chapter 1 for a discussion about ignorability). In practice, the ignorable assumption typically implies either MCAR or MAR mechanism provided that the parameters describing the missing data mechanism are distinct from the parameters of model of interest (see also Verbeke and Molenberghs, 2000).

> → *Hence, the validity of the random-effects or marginal methods highly depends on the assumption that the missingness mechanism can be ignored.*

We note that when missing observations are limited to the dependent variable only, the DL approach implies applying random-effects models. Since the data are incomplete, subjects do not have necessarily equal contributions to the parameter estimation process. For instance, a subject with four measurements has more contribution than a subject with three measurements. As a result, the likelihood function that is maximised in the random-effects models is indeed the observed likelihood function that is constructed by the DL approach.

Let us revisit the Growth study, wherein the growth measurements of 11 girls and 16 boys were recorded at four ages. The original data contain no missing observations, but Little and Rubin (1987, p. 159) generated an incomplete version by removing nine (out of 27) measurements at age 10. Following Little and Rubin (1987), it is more likely for children with a low growth score at age 8 to have missing observations at age 10. Unlike real settings, the process in which the missing data are generated is known. So, we know that the process of missing data generation depends on the observed values (at age 8) and not on the missing values themselves. Hence, it can be concluded that the missing data mechanism is MAR. Because the missing observations are subject to the dependent variable only (some measurements at age 10 are missing) and the missing data mechanism can be ignored due to the MAR assumption, the standard random-effects model is appropriate and delivers unbiased estimates. Table 3.10 shows the results of the random-intercept model fitted to the incomplete data.

As can be seen, these results are consistent with the results of the same model before introducing missing observations (in Table 3.1). Standard errors, however, are inflated due to missing observations. Of course, this is expected because the number of available measurements in the incomplete data is less than the number of measurements in the complete data. It is also interesting to evaluate the random components of the model, which are shown in Table 3.11. The between-subject variance ($\hat{\tau}^2 = 3.38$) and the within-subject variance ($\hat{\sigma}^2 = 2.08$) seem to have increased marginally in comparison to the analysis before introducing missing data. This results in a slightly smaller, though still comparable, intra-class correlation.

Hence, the random-effects models provide valid estimates in the analysis of longitudinal data with missing observations. The main reason for validity of the method is that it uses all available information without discarding any incomplete cases from the analysis, and henceforth is superior to CCA. Moreover, any kind of imputation procedures, such as LOCF or even

TABLE 3.10

Growth Study: Estimates of the Fixed Effects

Parameter	Estimate	Std. Error	df	t-value	p-value	95% Confidence Interval Lower Bound	Upper Bound
Intercept	16.30	1.03	94.56	15.82	0.000	14.25	18.35
Sex	0.92	1.62	94.54	0.57	0.572	−2.29	4.13
Age	0.79	0.08	70.00	9.68	0.000	0.62	0.95
Sex × Age	−0.30	0.13	70.00	−2.33	0.023	−0.55	−0.04

Dependent Variable: *Distance*.

TABLE 3.11

Growth Study: Estimates of the Random Components

Parameter		Estimate	Std. Error	ICC
Repeated Measures	Variance $\hat{\sigma}^2$	2.08	0.35	
Intercept	Variance $\hat{\tau}^2$	3.38	1.12	0.62

Dependent Variable: *Distance*.

multiple imputation, is redundant in such situations. There are also no additional steps involved in using the random-effects model in comparison to, for instance, multiple imputation, wherein an additional step is required to create multiple copies of the completed datasets (i.e. imputed datasets).

In conclusion, (linear) random-effects models are preferable to the other methods including multiple imputation based upon two conditions:

1) Missing values are confined to the dependent variable.

2) Missing data mechanism can be ignored.

Although random-effects models are the preferred choice when missing observations are confined to the dependent variable, these models rely heavily on the MAR assumption. Exceptionally, it was known to us how the missing observations were deliberately introduced in the Growth study. We could thus argue the reasons for missing data and justify the MAR assumption. In reality, however, subjects often leave the study prematurely due to reasons that might be related to their (unobserved) measurements of the dependent variable. In such cases, it would be very difficult to verify the MAR assumption. Consequently, it is always advisable to investigate how plausible the MAR assumption is for a given dataset. This could be best achieved by trying to obtain information from external sources. This can, for example, be done by contacting subjects who dropped out from the study to verify the reason(s) for dropout.

Furthermore, because the MAR assumption cannot generally be justified from the observed data, it is recommended to apply models that go beyond the MAR assumption and allow for more general missing data mechanisms. Moreover, there might be cases for which missingness is related to the unobserved dependent variable, so that the missing data mechanism can no longer be ignored. These imply that the missing data mechanism should be explicitly taken into account in the analysis phase. In the following section, we review two classical approaches for nonignorable models, though alternatives exist (see, for instance, Fitzmaurice et al. 2012, Chapter 19).

Consider the random-effects model defined in Equation (3.23) in the Growth study, that is,

$$Distance_{ij} = \beta_0 + \beta_1\, Age_{ij} + V_{ij} \qquad (3.23)$$

for child i at time j. This model is the analysis model of interest. Let us also define an indicator variable I_{ij} such that $I_{ij} = 1$ if the outcome of child i is observed at time j. Otherwise, $I_{ij} = 0$ meaning that the intended measurement is not observed for that child at time j. To keep the presentation simple, we assume the process governing the missingness follows the logistic regression model

$$logit\left\{Pr\left(I_{ij} = 1\right)\right\} = \gamma_0 + \gamma_1 Age_{ij} + \gamma_2 Distance_{ij} + +\gamma_3 Distance_{i(j-1)}, \qquad (3.24)$$

where logit transformation is
$logit\{Pr\ (I_{ij} = 1)\} = log\ \{Pr\ (I_{ij} = 1)/Pr\ (I_{ij} = 0)\}$. Equation 3.24 implies the probability of observing the dependent variable measurement (i.e., *Distance*) for child i at time j depends on the age of the child at time j, measurements of the outcome at time j and $(j\text{-}1)$. For example, for a child with a missing dependent variable at time 2, Equation 3.24 allows dependency on *Age* at 10, *Distance* at age 10 (time 2), and *Distance* at age 8 (time 1). Note that this model could additionally allow for dependency on the future outcomes, on other previous outcomes, or on external independent variables.

Equations 3.23 and 3.24 form together a joint statistical model that accommodates the missing data mechanism into the modelling process. It is known as the *selection model* in the literature (see, among others, Ibrahim and Molenberghs, 2009). A big challenge is that the relation between the primary analysis of interest and missing data process cannot be derived from the observed data. In a monotone pattern, this simply implies that γ_2 is inestimable in Equation 3.24. Intuitively, it can be argued that γ_2 cannot be estimated because $Distance_{ij}$ is not fully observed in Equation 3.24. Of course, if we impose some assumptions about unobserved $Distance_{ij}$, and hence about all components of $Distance_{ij}$ (e.g., normality of the full vector $Distance_{ij}$), this parameter can be estimated. Nonetheless, such analyses heavily depend on the validity of these assumptions. It is therefore important to realise that any conclusion from a nonignorable model (similar to ignorable models) is sensitive to the posited assumptions about the missing data. Hence, the analysis of missing data should include an investigation of how robust the results are with respect to deviations from these assumptions that cannot be verified from the observed data. As an example, a t-distribution assumption can be considered for $Distance_{ij}$ to investigate how sensitive the results are with respect to the normality assumption.

→ *We thus define a sensitivity analysis as a set of analyses to evaluate the robustness of the results to the unverifiable assumptions about missing observations.*

Although construction of selection models is fairly straightforward, the estimation process of these models (based on the DL approach) is seriously complicated, as it involves evaluations of multiple integrals. Unfortunately, standard statistical packages such as SPSS do not offer this option yet, and so, the analysis of such models is currently lacking in the main-stream statistics software. For relatively simple settings, however, it is still possible to apply an approximately equivalent approach, known as the two-stage Heckman approach, to address the problem of missing not at random data using standard statistical packages. In the accompanying website, we will illustrate how the selection model can be fitted (using SPSS) to a pre-post design when the missing data mechanism is nonignorable.

An alternative taxonomy of nonignorable models can be built based on the so-called pattern-mixture framework (Little, 1994), where distinct models are defined for each pattern of the observed data. In the Growth study, for instance, the data consist of the following two patterns:

Pattern 1: Children who have all four measurements (and thus complete)

Pattern 2: Children who have measurements at age 8, 12 and 14 (and thus missing at age 10).

Note that pattern 1 consists of children who have all measurements, while patten 2 includes those nine children who do not have the measurement at age 10.

By analogy with model Equation 3.23, the models can be expressed as

$$Distance_{ij} = \beta_0^{(p)} + \beta_1^{(p)} Age_{ij} + V_{ij}^{(p)}, \tag{3.25}$$

for patterns $p = 1, 2$. We should emphasise that each model has its own parameters. For example, $\beta_0^{(1)}$ and $\beta_1^{(1)}$ are the fixed-effects parameters in pattern 1, whereas $\beta_0^{(2)}$ and $\beta_0^{(2)}$ are the fixed-effects parameters in pattern 2. Likewise, $V_{ij}^{(1)}$ and $V_{ij}^{(2)}$ are different variance-covariance components in patterns 1 and 2, respectively.

In the above pattern-mixture models, some parameters are unidentifiable. Specifically, the first pattern consists of four measurements (*Distance* at *Age* 8, 10, 12 and 14), whereas only three measurements are available in the second pattern (i.e., *Distance* at *Age* 8, 12 and 14). As a result, all parameters related to *Age* = 10 in the second pattern are inestimable. For the abovementioned models, this can be best seen in the estimate of the random components. Table 3.12 shows the estimate of $V_{ij}^{(1)}$ and $V_{ij}^{(2)}$ after fitting Equation 3.25 to each pattern separately. As can be seen, no estimate is available for *Age* at 10 in the second pattern (see part b).

TABLE 3.12

Growth Study: Estimate of the Variance-Covariance Matrix with a Compound Symmetry Structure for Patterns 1 and 2, Respectively

Residual Variance-covariance (V) Matrix[a]

	$[Age = 8]$	$[Age = 10]$	$[Age = 12]$	$[Age = 14]$
a)				
$[Age = 8]$	5.45	3.59	3.59	3.59
$[Age = 10]$	3.59	5.45	3.59	3.59
$[Age = 12]$	3.59	3.59	5.45	3.59
$[Age = 14]$	3.59	3.59	3.59	5.45

[a] Dependent Variable: *Distance*.

Residual Variance-covariance (V) Matrix[a]

	$[Age = 8]$	$[Age = 12]$	$[Age = 14]$
b)			
$[Age = 8]$	5.98	2.91	2.91
$[Age = 12]$	2.91	5.98	2.91
$[Age = 14]$	2.91	2.91	5.98

[a] Dependent Variable: *Distance*.

Compared with the selection models, an advantage of the pattern-mixture approach is that it makes explicit which parameters cannot be identified and therefore forces to impose restrictions on unidentified parameters, which in turn, simplifies the task of performing sensitivity analyses. The key advantage of pattern-mixture models is therefore transparency about verifiable and unverifiable assumptions.

However, these models do not directly provide marginal quantities of substantive interest like the overall treatment effect or time evolution. In the Growth study, the relationship between *Distance* and *Age* is evaluated in subgroups defined by patterns, so the overall effect of age is unavailable directly from the pattern-mixture analyses. In order to obtain the overall profile for the whole study, estimates from different patterns are combined by averaging across patterns. Let π_p, $p = 1,...,P$, denote the population percentage of subjects for pattern p. The overall estimate of interest (e.g. the fixed effect β_1) is then a weighted sum of the estimates, that is,

$$\beta_1 = \pi_1\beta_1^{(1)} + \cdots + \pi_P\beta_1^{(P)}, \quad (3.26)$$

with population weights replaced by the proportion of cases in the data. In a similar fashion, with a minor adjustment, the variance of the overall estimate can be obtained (see, Hedeker and Gibbons, 2006, p. 309). In Chapter 6, we will illustrate a pattern-mixture model (by imposing restrictions on unidentified parameters) using the existing mixed procedure in SPSS.

A final note is that both selection and pattern-mixture models rely on untestable assumptions (i.e. some parameters are inestimable). In selection models, these assumptions are implicit, whereas they are explicit in the pattern-mixture models. As indicated in the study by Verbeke and Molenberghs (2000, pp. 217–219), the main concern should not be centred on whether the selection model or the pattern-mixture model should be used in practice (as each has its useful features). The focus instead should be on evaluations of unverifiable assumptions by means of sensitivity analyses. Furthermore, it should be emphasised that the MAR assumption itself is also untestable. Given MAR, the random-effects models can be used, but the MAR assumption itself cannot be tested. This fact is often overlooked when these models are used in practice.

3.5.4.2 Missing Observations in the Independent Variables Only

Sometimes, missing observations are limited to the independent variable(s) while the dependent variable is fully observed. Specifically, this may happen when the baseline measurements have missing data. For example, Everitt (1994) reported the results of a 100-kilometre (km) road race in the UK, where the running time of every runner is reported in each 10 km section of the race together with the age of the runner. Since all runners finished the race, the dependent variable was fully measured. However, the age was not reported for some runners leading to a longitudinal data with missing observations only at the baseline variable (i.e. age). Random-effects models do not automatically allow incompleteness in the baseline or time-varying independent variables, and hence, an analysis of longitudinal data in the presence of missing observations in those variables requires exclusion of subjects for whom the independent variables are not fully observed. Here, the parameter estimates can be biased if the missingness depends on the dependent variable. Tan et al. (2018) demonstrated this by means of a simulation study when the time-varying independent variables were incompletely observed. Another issue with such exclusions is that they are wasteful of information, as those subjects with missing observations are removed from the analysis while their dependent variable might still be observed.

A better strategy is to impute missing observations of the independent variables and reproduce the complete matrix. When an imputation methodology is contemplated, MI is recommended because it correctly reflects uncertainty about which value to impute. Moreover, this option is currently widely available in the main-stream statistical software. In Chapter 4, we discuss different ways of imputing missing observations in longitudinal studies.

Although MI is popular and readily available, it can deliver undesirable results if the missing observations are imputed incorrectly, that is, the estimates can be biased if the model used for drawing imputations (i.e. imputation model) is incorrect or mis-specified. Hence, special attention to specify a (nearly) correct imputation model is needed when MI is used. With a low percentage of missing observations (in the independent variables), the effect of a possible mis-specified imputation model could be minimal. However, when the percentage of missing observations becomes larger, a great deal of attention should be given to the specification of the imputation model. In Chapter 4, we elaborate on these points and provide practical recommendations on defining appropriate imputation models in longitudinal settings.

3.5.4.3 Missing Observations in the Dependent and Independent Variables

In most applications, the dependent, baseline or time-varying variables may not be fully observed for all subjects leading to missing data in both the dependent and independent variable(s). As discussed earlier, subjects with incomplete independent variables should not be excluded, and so, the choice of preference is MI. It is thus more convenient to apply MI to all variables simultaneously. Moreover, additional information such as secondary dependent variable or auxiliary variables that are not a part of the main analysis can be incorporated into the imputation process to improve imputations so as to make the MAR assumption more plausible. Chapter 4 discusses a detailed example of MI for handling missing observations.

It is sometimes advocated to first impute missing observations of the independent variables, but not dependent variable, and then apply random-effects models to the data. This, however, is challenging because it is unclear how those missing observations can be imputed if their corresponding dependent variable has missing observations as well. In the Growth study, for instance, suppose *Sex* of some children is missing. For them, if one or more distance scores (i.e. dependent variable) are also missing, these missing distance scores do not contribute to the imputation of *Sex*. Consequently, the imputed values of *Sex* are imprecise and may cause biased results.

There are certain situations in which imputation of missing observations in the baseline variables is useful. Following Kayembe et al. (2020), mean imputation of baseline variables in randomised controlled trials is valid (i.e. results in unbiased treatment effect estimates and standard errors). In such exceptional situations, it is therefore more convenient to first impute the missing baseline variables deterministically and then to apply random-effects models to the incomplete data where the dependent variable has missing observations. Of course, the corresponding results are valid as long as the missing data mechanism for dependent variable can be ignored. When time-varying independent variables have missing observations, deterministic imputation approaches are not typically recommended, so MI should be used.

3.5.5 Conclusion and Recommendations

This section summarizes the results of using different methods for handling missing data issues in longitudinal studies. In particular, it has been discussed that CCA delivers unbiased estimates of parameters in multilevel models when the reason(s) for missing data are unrelated to the dependent variable at any time point. If it turns out that some variables (that are not necessarily part of the main analysis) are related to the missingness, inclusion of such variables in the analysis makes CCA valid. Hence, CCA is only recommended when the dependent variable is not the cause of missingness. It must, however, be emphasised that CCA uses less information and so estimates are subject to potential loss of efficiency.

When the missing data mechanism depends on the observed part of the dependent variable, the DL or MI approach is recommended. For missing observations in the dependent variable, the random-effects or marginal models are easier to implement, while MI is more straightforward with missing observations in the independent variables (or both). These state-of-the-art methods, however, should be extended to allow for nonignorable missingness when the unobserved part of the dependent variable is the cause of missing observations, and sensitivity analysis should be carried out to assess the degree to which the conclusions vary across ranges of plausible scenarios. Table 3.13 summarises the minimally appropriate method when dealing with missing observations in longitudinal studies. Similar to Chapter 2,

TABLE 3.13

The Suitable Methods to Handle Missing Data in Longitudinal Studies

| | | Missing Data Mechanism | | |
| | | Depends on | | Does Not Depend on |
Missing Data		Unobserved Dependent Variable	Observed Dependent Variable	Dependent Variable at All
location	Dependent variable	DL/MI + SA	DL/MI (+SA)	CCA
	Independent variable(s)	--	DL/MI (+SA)	CCA
	Both	DL/MI + SA	DL/MI (+SA)	CCA

Methods are complete-case analysis (CCA), direct likelihood (DL), multiple imputation (MI), and sensitivity analysis (SA).

we emphasise that the DL or MI approach can be considered as the starting point for conducting sensitivity analysis when the missingness mechanism depends on the dependent variable. This practice can be optional when the missingness mechanism depends on the observed dependent variable only (second column).

Finally, it is worth mentioning that the observed data provide no information to either confirm or refute dependency of missingness on the observed or unobserved dependent variable. For this reason, reliance on a single model to make inference for a given dataset is not advisable. Instead, one may start with a standard approach (i.e. DL or MI) and move towards sensitivity analysis by choosing a set of different assumptions about the nature of missingness.

Here, MI is a useful tool in developing sensitivity analysis strategies, as it can be combined with pattern-mixture approach since it offers computational simplicity in some situations as compared to selection models. We provide illustrative examples for conducting sensitivity analyses in Chapters 5–7.

3.6 Assignments

3.6.1 Assignment

Consider the study about the relationship between alcohol consumption and violent behaviour (see section 'Short description of research and simulation study' and Section 2.3.1 for a description of the study) (SPSS system file: Alc_violent.sav).

Analyse the relationship between alcohol consumption and violent behaviour, address the following sub-questions:

a. Perform a random-effects model analysis with a random-intercept model to study the subject-specific relationship between alcohol consumption and violent behaviour and compare this output with that of assignment 2.8.13.

b. Determine the V_{cov} matrix and V_{corr} matrix and interpret the ICC of the model obtained in a.

c. Can you generalise your findings to a larger group of subjects?

d. Visualise concepts such as subject-specific line and within- and between-subjects variance by means of plots (constructed by hand).

3.6.2 Assignment

Growth data (Pothoff & Roy: see section 'Short description of research and simulation study' and Section 3.1.2 for a description of the study) (SPSS

system file: Growthdata.sav). Consider the study about the orthodontic growth of boys and girls. Compare growth and growth velocity between boys and girls, address the following sub-questions:
Perform a random-intercept model

$$Distance_{ij} = \beta_{0i} + \beta_1 Age_{ij} + \beta_2 Sex_i + \beta_3 Age_{ij} \times Sex_i + R_{ij}$$

Questions:

a. Is the interaction between *Age* and *Sex* significant?
b. Explain the discrepancy with assignment 2.8.12 regarding the s.e.'s of b_3.
c. What is the interpretation of the first-level variance ($\mathbf{R}_{cov}$ matrix), the second-level variance ($\mathbf{G}_{cov}$ matrix) and the overall variances – covariances ($\mathbf{V}_{cov}$ matrix)?
d. Determine the $\mathbf{V}_{cov}$ matrix and $\mathbf{V}_{corr}$ matrix and interpret the ICC?
e. Visualise concepts such as subject-specific line and within- and between-subjects variance by means of plots (constructed by hand).

3.6.3 Assignment

Consider the Growth study (SPSS system file: Growthdata.sav) and the following marginal model

$$Distance_{ij} = \beta_{0i} + \beta_1 Age_{ij} + \beta_2 Sex_i + \beta_3 Age_{ij} \times Sex_i + V_{ij}.$$

Perform a marginal model with the (homogeneous) 'Compound symmetry' covariance structure.

a. Compare the variances and covariances of the output with that of the
b. random-intercept model in assignment 3.6.2.
c. Perform a random-intercept/slope (random slope for age) model of the Growth study. Compare all the output and argue which model you would prefer.
d. Determine the $\mathbf{V}_{cov}$ –and $\mathbf{V}_{corr}$ –matrix of the models in a and b.

3.6.4 Assignment

Consider the Salsolinol study (SPSS system file: Salsolinol.sav). See section "Short description of research and simulation study" and Section 4.2.3 for a description of the study.

a. Plot the subject-specific and the average *TSALS* against the time points for subjects with moderate and severe dependency on alcohol. Describe what you see and which systematic part of the model would you specify?

b. Perform a random-intercept subject-specific model and a marginal model with (homogeneous) "compound symmetry." Why doesn't the random-intercept model give valid estimates?

3.6.5 Assignment

Consider the random-intercept/random (RI/RS) slope model Equation 3.10:

$$Distance_{ij} = \beta_{0i} + \beta_{1i} Age_{ij} + R_{ij},$$

where

$$\beta_{0i} = \beta_0 + G_{0i}, \text{and}$$
$$\beta_{1i} = \beta_1 + G_{1i}$$

a. Show that Equation 3.10 can be written as a marginal model:

$$Distance_{ij} = \beta_0 + \beta_1 Age_{ij} + V_{ij},$$

where

$$V_{ij} = G_{0i} + G_{1i} Age_{ij} + R_{ij}$$

b. Suppose that *Age* is considered nominal. The values are 8, 10, 12, and 14 years. What is the specification of the RI/RS model?

c. What is the specification of the RI/RS model for the Proximity study?

3.6.6 Assignment

Consider the Proximity study (SPSS system file: Teacher.sav). For a short description, see section "Short description of research and simulation study" and Section 2.4.1.

a. Plot the teacher-specific proximity score versus occasion (use the variable "*Occ*") for each *Sex* and a plot of the sex-specific proximity score versus *Occ*. What can you say about the (Teacher specific/Sex specific) pattern of proximity score across occasions?

b. Argue that the following model specification does make sense.

$$Proximity_{ij} = \beta_0 + \beta_1 Occ_{0ij} + \beta_2 Occ_{1ij} + \beta_3 Occ_{2ij} + \beta_4 Sex_i$$
$$+ \beta_5 Occ_{0ij} \times Sex_i + \beta_6 Occ_{1ij} \times Sex_i + \beta_7 Occ_{2ij} \times Sex_i$$
$$+ random\ part + R_{ij}$$

The variable Occ_{ij} denotes the dummy variable for subject i on occasion $j = 0,1,2$.

c. Can you make an educated guess whether a random intercept or a random slope (with random intercept) would be most appropriate to describe the data?

d. Perform successively with the same systematic part as in b

   i OLS model

   ii A random-intercept model

   iii A random-intercept/slope model

   iv A random-intercept/slope model with an AR (1) serial correlation and homogeneous variances.

   v A random-intercept model with AR (1) and heterogeneous variances and

   Which model would you consider as best and why?

e. Would you conclude that there is a difference in change of proximity score between male and female teachers?

3.6.7 Assignment

Consider the Alzheimer study that is briefly introduced in Section 3.5.1 (SPSS system file: Alzheimer.sav).

a. Obtain the patterns of missing data. Are the patterns monotone or non-monotone?

b. How many patterns exist in this study? Produce a table with the percentage of missing observations per pattern. Which pattern has the highest frequency?

c. Investigate whether the patterns of missing data depend on time, treatment and the interaction between time and treatment.

d. Consider the following marginal model

$$Score_{ij} =$$
$$\beta_{0i} + \beta_1 week_{ij} + \beta_2 treatment_i + \beta_3 week_{ij} \times treatment_i + V_{ij}.$$

with a compound symmetry covariance structure. For this model

    i Perform the complete-case analysis.

    ii Perform the direct likelihood analysis.

    iii Compare the results of the above analyses (in parts i and ii). What can you conclude with respect to the missing data mechanism?

3.6.8 Assignment

In the Proximity study (see section "Short description of research and simulation study" and Section 2.4.1 for a short description), suppose the model of interest is a random-intercept model with an AR(1) and heterogeneous variances:

$$Proximity_{ij} = \beta_{0i} + \beta_1 Occ_{0ij} + \beta_2 Occ_{1ij} + \beta_3 Occ_{2ij} + \beta_4 Sex_i + \beta_5 Occ_{0ij}$$
$$\times Sex_i + \beta_6 Occ_{1ij} \times Sex_i + \beta_7 Occ_{2ij} \times Sex_i + R_{ij}$$

The variable Occ_{kij} denotes the dummy variable Occ_k for subject i on occasion $j = 0,1,2$. $k = 0, 1, 2$ (SPSS system file: Teacher.sav). For this model,

a. Perform the direct likelihood analysis

b. Perform the complete case analysis

c. Impute missing observations using the LOCF approach and apply the above model to the completed (imputed) data

d. Compare the results of part a, b, and c.

4

Model Building for Longitudinal Data Analysis

4.1 Basic Guidelines

This section presents three basic steps that should be considered when dealing with the analysis of longitudinal data. Specifically, we have pointed out in Section 3.4 that there might exist several different competing variance-covariance structures in a longitudinal study and objective criteria are needed to decide which structure is the best.

We further assume that missing observations are limited to the dependent variable, if it exists. The issue of missing data is discussed in greater detail in Section 4.3. It should also be noted that the proposed marginal or random-effects models can lead to biased results if missingness in the dependent variable depends on the unobserved dependent variable (see Section 3.5.4).

4.1.1 Step 1. Exploratory Analysis

Before specifying the marginal or random-effects model, an exploratory analysis in light of the research question is usually the first step. On the one hand, an important aspect in the exploration is to investigate whether the planned fixed part of the model meets the linearity assumption between the independent variables and the dependent variable (see Section 2.5.1). Linearity assumption between *Time* and the dependent variable may also be relevant (see, e.g., the Growth study). On the other hand, linearity will not be an issue if *Time* is considered categorical (see, e.g., Proximity study). A drawback is that the number of parameters will increase with the number of time points.

Another important point is whether there are observations that disproportionately determine the trend of the regression lines. This will be the case if, for example, there are wrong codes or influential cases (see, e.g., Ouwens et al. (2001), Tan et al. (2001)). These are the checks that are standard in regression analysis.

Once the full relevant systematic part of the model is specified, the observed correlations of the residual scores between different time points and variances of each time point together with residual plots can be studied, which may lead to a well-educated guess about the variance-covariance structure of the data.

4.1.2 Step 2. Procedure for Choosing the Best Variance-Covariance Structure

If a correct variance-covariance structure $\mathbf{V}_{cov}$ is not specified for the analysis model, the estimates of the regression parameters may be biased (Gurka et al., 2011). Several different candidates of variance-covariance structures are encountered in longitudinal data. In this step, different variance-covariance structures will be compared with the aid of statistical tools to determine the simplest, yet still appropriate structure (without losing significant information). Specifically, we choose the best fitting structure (in the sense of a smaller number of parameters and without significant loss of information) by comparing various marginal and random-effects models using the loglikelihood ratio test for nested models or by using an information criterion such as BIC or AIC for non-nested models. In general,

> → *Model A is nested within model B (hierarchically related) if model A is obtained from model B by*
> - *removing variables, variance and/or covariance parameters or by*
> - *imposing restrictions with respect to the variance-covariance parameters of model B (see e.g., Widaman, 1985).*

If two competing variance-covariance structures S1 and S2 are not nested, but there is a third variance-covariance structure (S3) nested within S1 and S2 that fits equally well using the likelihood criterion, then S3 is considered to be the better choice than both S1 and S2.

4.1.3 Step 3. Interpretation of the Regression Parameter Estimates

After choosing the best variance-covariance structure, a backward procedure is performed for the fixed part of the model, while keeping the chosen variance-covariance structure. As a last step, the estimated regression parameters of the final model are interpreted.

In the next section, the Proximity study, the Growth study and the Salsolinol study serve as real-life examples to clarify the process of analysing multilevel data.

4.2 Analysis of Some Case Studies

4.2.1 Best Practice for the Analysis of Proximity Data

In the Proximity study described in Section 2.4.1, 36 male and 14 female teachers were evaluated on their proximity scores at four occasions. Suppose that the main interest lies in change of average proximity scores over time for male and female teachers separately. Thus, we plan to compare the average proximity scores between successive occasions. We therefore need to test the change in the population average proximity score (*Proximity*) from the first (Occ_0) to the second occasion (Occ_1), from the second (Occ_1) to the third occasion (Occ_2), and from the third (Occ_2) to the fourth occasion (Occ_3). Hence, six comparisons are planned in advance: three for females and three for males. Note that a comparison between male and female teachers per occasion is of no interest now (see assignment 4.3.3 for such a comparison).

4.2.1.1 Step 1. Exploratory Analysis

The data of Proximity study are sampled from a two-level design with teachers as the second-level and repeated measurements (within teachers) as the first-level. *Proximity* and *Occasion* are the first-level variables, while *Sex* is the second-level fixed variable. Because students fill in the same questionnaire every year, serial correlation may also exist due to a possible memory effect.

Following the study aim and discussion of Section 3.3, we define the following full relevant model:

$$Proximity_{ij} = \beta_0 + \beta_1 Occ_{1ij} + \beta_2 Occ_{2ij} + \beta_3 Occ_{3ij} + \beta_4 Sex_i + \beta_5 Occ_{1ij}$$
$$\times Sex_i + \beta_5 Occ_{2ij} \times Sex_i + \beta_5 Occ_{3ij} \times Sex_i + V_{ij}, \quad (4.1)$$

where the dependent variable $Proximity_{ij}$ is the proximity score of subject i at occasion $j = 0,\ldots, 3$. The variables of the systematic part are defined as in Equation 3.12. The next task is to make an educated guess about each element of the variance-covariance matrix $\mathbf{V}_{cov}$.

This can be done, for instance, by applying the standard OLS method to Equation 4.1. Note that the corresponding residuals must be transposed to the wide format such that they are presented column-wise per time point (see Section 4.3.1 for a comparison between the wide and long format of data). These can then be used to obtain descriptive statistics and pairwise correlations for the residuals, which are essentially correlated. These correlations are denoted as the observed correlations.

Table 4.1 shows descriptive statistics for the residuals at each occasion. A closer look reveals that the sample size drops from 46 at baseline (Occ_0) to 32 after three years (Occ_3) due to missing observations. Furthermore, the standard deviation of residuals indicates a decreasing trend (from 0.54 to 0.33), suggesting unequal variances across occasions. Additionally, Table 4.2 shows that the observed pairwise correlations (of the residuals) decrease as the time span between two observations increases.

From Tables 4.1 and 4.2 and Figure 3.7 (the subject-level profiles), several variance-covariance (correlation) structures can be inferred. A random-intercept (RI) structure would not be a good match because of the decreasing correlation over time, as the correlations are constant in the RI model (see Equations 3.6 and 3.7). In contrast, a random-intercept/random-slope (RI/RS) structure, or perhaps in combination with an AR (1) serial correlation structure (with equal variances) are more plausible candidates. The latter is based on substantive arguments that there may be some memory effects since students were asked to fill in the same questionnaire every year. Another candidate could be an AR (1) structure with heterogeneous variances because Tables 4.1 and 4.2 show unequal variances with decreasing correlations for the residuals. Finally, an unstructured (UN) variance-covariance structure that poses no constrains in the variance-covariance matrix $\mathbf{V}_{cov}$ can also be considered. Note that other plausible correlation structures or a user-specified $\mathbf{V}_{cov}$ (assuming the statistical package permits) might also be considered. In this book, we however restrict ourselves to the variance-covariance structures that are mentioned in Section 3.4.

TABLE 4.1

Proximity Study: Descriptive Statistics of the Residuals at each Occasion from the Standard OLS Method

	N	Minimum	Maximum	Mean	Std. Deviation
Residual Occ_0	46	−1.38	0.90	0.00	0.54
Residual Occ_1	38	−1.09	0.91	0.00	0.43
Residual Occ_2	37	−0.82	0.82	0.00	0.36
Residual Occ_3	32	−0.57	0.47	0.00	0.33

TABLE 4.2

Proximity Study: Observed Correlation Matrix ($\mathbf{V}_{corr-observed}$) based on the Residuals at each Occasion from the Standard OLS Method

Residual Occ_0		Residual Occ_1	Residual Occ_2	Residual Occ_3
Residual Occ_0	1	0.68 (35)	0.68 (33)	0.46 (27)
Residual Occ_1	0.68 (35)	1	0.64 (29)	0.61 (24)
Residual Occ_2	0.68 (33)	0.64 (29)	1	0.80 (23)
Residual Occ_3	0.46 (27)	0.61 (24)	0.80 (23)	1

Sample sizes are between bracket.

Furthermore, a scatter plot matrix of residuals (see Figure 4.1) can be used to investigate whether there are obvious extreme values or influential observations (i.e. those of having a disproportional effect on correlations). The scatter plot in the first row and second column shows the residuals at the first occasion against the residuals at the second occasion (after one year). Likewise, the scatter plot in the first row and third column shows the residuals at the first occasion against the residuals at the third occasion (after two years). For a discussion of how to detect influential observations and structures, see, for example, Ouwens et al. (2001) and Tan et al. (2001). From Figure 4.1, we observed no sign of extreme values or influential observations.

From this exploratory analysis, several **V** matrices are candidates for the Proximity study. Note that although the UN variance-covariance structure is the most general structure, as no restriction is placed on the variances and covariances, the number of parameters increases dramatically when the number of time points increases. For example, the UN variance-covariance structure requires estimation of 55 parameters with 10 time points. Consequently, simpler structures that fit the data equally well are preferred in practice. For equidistant time points, for instance, the Toeplitz with heterogeneous variances is close to UN, but it only requires an estimation of 19 parameters with 10 time points (i.e. 10 variances and 9 covariances). In the Proximity study, the UN and Toeplitz variance-covariance structures consist of 10 and 7 parameters, respectively. For non-equidistant time points, the ante-dependence structure with the same number of parameters would be an alternative to the UN variance-covariance structure.

The next step is to find an objective criterion to determine which variance-covariance structure among the chosen alternatives is the best choice.

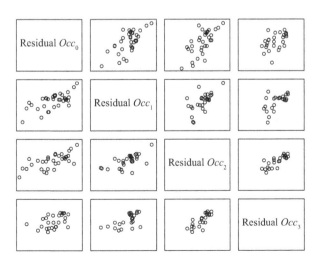

FIGURE 4.1
Scatter plot matrix of residuals in the Proximity study using the standard OLS.

4.2.1.2 Step 2. Procedure for Choosing the Best Variance-Covariance Structure

For hierarchical (nested) models, the 'Restricted -2LL' (log likelihood) can be used to construct the well-known likelihood ratio test to compare alternative models that only differ in variance-covariance structures. The term *restricted* (based on the REML estimates) is added to distinguish it from the commonly used -2LL test criterion (based on the ML estimates). The ML estimates of the variances and covariances lead to underestimation of the standard errors of the estimated variances and covariances and are therefore not recommended in this phase of the analysis (see Verbeke & Molenberghs, 2000, Section 5.3).

Roughly speaking, the restricted -2LL test criterion is a distance measure between the fitted model and the observed data. As a general rule, the smaller the restricted -2LL, the better the model fits to the data. Consequently, when comparing two hierarchical (or nested) models that only differ in variance-covariance structure, the difference in their restricted -2LL will be calculated, which is called the restricted log-likelihood ratio. According to a statistical theory, this difference has (asymptotically) a Chi-square distribution with known degrees of freedom. The calculation of the degrees of freedom, however, is not always straightforward in certain cases, for example when comparing an RI with an RI/RS model. For a thorough discussion, we refer to studies by Verbeke and Molenberghs (2000), Fitzmaurice, Laird, and Ware (2012), and Snijders and Bosker (2012).

For ease of presentation, we propose the following procedure to test two competing models (see also Snijders and Bosker, 2012):

1. The two-competing hierarchical (i.e. nested) models must have the same fixed-effects part.
2. Estimate all parameters with REML.
3. Calculate the difference in restricted-2LL between the two models (restricted log-likelihood ratio test).
4. If the test concerns the significance of covariances, then the log-likelihood ratio test should be considered as a two-sided test with degrees of freedom equal to the difference in the number of model parameters.
5. If the test concerns the significance of variances, then the restricted log-likelihood ratio test should be used as a **one**-sided test with degrees of freedom equal to the difference in the number of model parameters.

Table 4.3 shows some common variance-covariance structures in multilevel models. Each column (or row), represented by a number, belongs to a variance-covariance structure. For example, number 5 represents the compound

TABLE 4.3

An Overview of Several (non-)nested Variance-Covariance Structures

	1	2	3	4	5	6	7	8	9	10	11	12
1 ind	X	<	<	<	<	<	<	<	<	<	<	<
2 RI(ho*)		X	<	++	++	<	NN	NN	<	<	NN	<
3 RI(he*)			X	+	++	NN	NN	NN	NN	<	NN	<
4 CS (ho)				X	<	++	NN	NN	<	<	NN	<
5 CS (he)					X	NN	NN	NN	NN	<	NN	<
6 RI/RS						X	NN	NN	NN	NN	NN	<
7 AR(1) (ho)							X	<	<	<	<	<
8 AR(1)(he)								X	NN	<	<	<
9 Toe (ho)									X	<	NN	<
10 Toe (he)										X	NN	<
11 Antedependence											X	<
12 UN												X

* ho: homogeneous variances, he: heterogeneous variances
A < B: A is nested within B (A > B: B is nested within A)
NN: non-nested variance-covariance structure
++: $\leq$ if correlations are positive and NN otherwise,
+: > if correlations are positive and NN otherwise.

symmetry structure with heterogeneous variances (CS (he)). In each row, a comparison is made between the corresponding variance-covariance structure and all variance-covariance structures after the 'X' sign. As an example, the CS (he) variance-covariance structure in row 5 is compared with RI/RS, AR (1) (ho) , ..., and UN variance-covariance structure. The sign '<' indicates nested structures, while the sign 'NN' means structures are non-nested. For instance, all variance-covariance structures are nested within the UN variance-covariance structure since imposing restrictions on the parameters of UN leads to the other variance-covariance structures. As opposed, on the one hand, the RI/RS structure and Toeplitz (homogeneous) are not nested structures and thus cannot be compared to each other using the likelihood ratio test. This is the case, because both structures have the same number of parameters (as we will see later in Table 4.9). On the other hand, the homogeneous (heterogeneous) AR (1) is nested within the homogeneous (heterogeneous) Toeplitz structure by imposing the restriction $\rho_2 = \rho_1^2$ and $\rho_3 = \rho_1^3$. Yet another example, suppose we want to compare two competitive models RI/RS and RI. The latter is nested within the former because the RI model can be obtained from the RI/RS model by imposing restrictions to the random slope variance and the covariance parameter τ_{01}.

For the Proximity study, a comparison of restricted -2LL for a number of variance-covariance structures is shown in Table 4.4. Recall that the UN, AR (1)

TABLE 4.4

Proximity Study: Alternative Variance-Covariance Structures, Ordered According to the Number of Parameters

	Model	Nr. of Estimated Parameters	-2LL (REML) BIC and AIC Resp. between Brackets
1	UN	18	112.88 (162.64, 132.87)
2	AR (1) heterogeneous variances	13	**120.23** (145.11, 130.23)
3	Random intercept + slope	12	**122.82** (142.73, 130.82)
4	Random intercept	10	137.66 (147.61, 141.66)
5	AR (1) homogeneous	10	137.54 (147.49, 141.54)
6	Independent (constant variances)	9	193.70 (198.68, 195.70)

with heterogeneous variances, and RI/RS variance-covariances structures are plausible candidates based on the exploratory analysis. Moreover, three additional alternatives, namely the independence, AR (1) with homogenous variances and RI structures, are added for the purpose of illustration. Note that both AR (1) structures target the possibility of serial correlations on top of correlation induced by subject differences, and the independence structure is included as an alternative structure to show that there is a significant amount of between-subjects variation that needs to be taken into account.

For the RI model (model 4), the -2LL value is 137.66, and 10 parameters were estimated (eight parameters for the fixed part in equation 4.1 plus one parameter for the random intercept variance τ_0^2, and one for the residual variance σ^2). The -2LL value of the RI/RS model (model 3) is 122.82, which has the same fixed part as given in Equation 4.1. Comparing these two models leads to a log-likelihood ratio value of 14.84 (= 137.66–122.82) with 2 degrees of freedom (i.e. the difference between the number of parameters in both models). It is important to emphasise that the actual degrees of freedom are slightly less than 2, and hence the proposed test tends to favour the more complex model than when the correct degrees of freedom are used (see, Fitzmaurice, Laird, and Ware, 2012, p. 209, and Snijders and Bosker, 2012).

The test statistics, that is the difference in -2LL of the RI and RI/RS models, can be used to test the null hypothesis.

H_0: RI structure, against the alternative

H_a: RI/RS structure.

The RI model implies that the slopes are the same for all subjects. Consequently, there is no random slope (i.e., $\tau_1^2 = 0$) and no covariance between the random slope and the random intercept ($\tau_{01} = 0$). Thus, the hypotheses can be reformulated as

$H_0: \tau_1^2 = 0$, against the alternative

$H_a: \tau_1^2 > 0$

A one-sided test should thus be used for the abovementioned hypotheses because the variance τ_1^2 cannot be negative. The corresponding one-sided p-value is equal to 0.000, which highly favours H_a. This confirms our expectation from Figure 3.7 that the RI/RS model is a better fit to the data than the RI model. Compared with the RI/RS model (-2LL = 122.82), the AR (1) heterogeneous model (model 2 in Table 4.4) has a lower -2LL of 120.23. Unfortunately, no comparison can be made because these competing models are not nested (see Table 4.3).

The Bayesian information criterion (BIC) is a popular criterion for selecting models that are not nested. It is a likelihood-based measure with a penalty term that depends on the number of parameters and the sample size. Unfortunately, there is still debate in the literature about the validity of BIC. Basically, the BIC is the logarithm of the likelihood value plus a penalty term that increases with the number of model parameters and the sample size. However, in some statistical packages (such as SPSS 27), this penalty does not take the correlation between repeated measurements into account and thus exaggerates the penalty (for a thorough discussion, see Delattre et al., 2014). Furthermore, a major drawback of BIC is that, in practice, it tends to favour simple models with a few numbers of parameters. For a discussion, the reader is referred to Weakliem (1999). In Section 4.2.3, we will demonstrate, through an example, that the BIC sometimes chooses an oversimplified model.

Like the BIC, the Akaike Information Criterion (AIC) is another information criterion suitable for the comparison of non-nested models. The difference is that BIC penalises model complexity more heavily.

→ *Vrieze (2012) made a comparison between AIC and BIC and concluded that AIC may be the preferred criterion, if the true model is not among the models.*

Unfortunately, in most situations, one cannot know whether the true model is among the models to be compared. Moreover, the results are based on a limited simulation study and no formal mathematical theory is available. To be on the safe side, in this book, we use both the criteria.

Finally, the common rule is to prefer the log-likelihood ratio test for nested models because the asymptotic distribution of the test statistic is known to be a chi-square distribution and thus the p-value is known. Nevertheless, substantive consideration prevails over statistical testing in general, particularly when the statistical theory is not fully available. In the Proximity study, the choice of alternative structures was partly based on subjective arguments. For example, the AR (1) was considered because there is a possibility of serial correlation.

After obtaining the restricted -2LL, BIC and AIC for the candidates of the variance-covariance structure in the Proximity study (Table 4.4), a strategy to determine the best variance-covariance structure works as follows.

The row and column of Table 4.5 are constructed such that the candidates are ordered according to a decreasing number of parameters. This table is then completed using Table 4.4 and the values of the likelihood ratio test together with the corresponding *p*-values. To compare the variance-covariance structures, consider the model displayed in a column and specify its variance-covariance structure as null-hypothesis. The variance-covariance structure of the models in the rows can be specified in the alternative hypothesis. For example, to compare model 2 (column-wise) with model 1 (row-wise), the following hypotheses can be formulated.

H_0: AR (1) heterogeneous, against

H_a: UN

We remind the readers that the above comparison is only possible because the AR (1) heterogeneous structure is nested within the UN variance-covariance structure (see Table 4.3). The two-sided likelihood ratio test leads to a *p*-value equal to $p = 0.20$. Thus, the AR (1) heterogeneous structure in the null

TABLE 4.5

Proximity Study: The Scheme to Determine the Best Variance-Covariance Structure

	Model	Par	1	2	3	4	5	6
1	UN	18	X	Df=5 Chi2=7.35 p=0.20	Df=6 Chi2=9.95 p=0.13	Df=8 Chi2= 24.78 p=0.00	Df=8 Chi2= 24.66 p=0.00	Df=9 Chi2= 80.83 p=0.00
2	AR (1) (he)	13		X	NP	NP	Df=3 Chi2=17.31 p=0.00	Df=4 Chi2=72.94 p=0.00
3	RI/RS	12			X	Df=2 Chi2=14.84 p=0.00*	NP	Df=3 Chi2=70.88 p=0.00*
4	RI	10				X	NP	Df=1 Chi2=56.04 p=0.00*
5	AR (1) (ho)	10					X	Df=1 Chi2=56.16 p=0.00
6	Ind	9						X

NP: no likelihood ratio test available.

* one sided.

hypothesis is preferred over the UN variance-covariance structure. Starting from the first row, it appears that only models 2 and 3 fit better than model 1 to the data, while the other models are not as good as the UN variance-covariance structure. Consequently, we do not have to compare the other models (model 4–6).

Model 2 (AR (1) heterogeneous) and model 3 (RI/RS) are, however, not hierarchically related so that these models cannot be compared using the likelihood ratio. These two models can therefore be compared using the BIC. Because model 3 has a somewhat smaller BIC, the RI/RS model can be preferred. The BIC for the RI/RS model is equal to 142.73, whereas the BIC for the AR (1) heterogeneous model is equal to 145.11. Unfortunately, the BIC does not provide a *p*-value (because the distribution is unknown) so that it is not clear whether the RI/RS model is significantly better than the AR (1) heterogeneous model. From a substantive point of view, the AR (1) heterogeneous model does make sense because it directly accounts for possible serial correlations. On the one hand, the RI/RS model may be preferred based on the BIC, where its variance-covariance structure has one parameter less. Using the AIC, on the other hand, the AR (1) heterogeneous model may be preferred. Since the purpose of the analysis is to estimate the regression parameters of the changes over time for each *Sex* separately, it makes sense to compare the regression estimates obtained from the RI/RS and AR (1) heterogeneous structures to investigate whether they differ (see assignment 4.5.3).

The procedure, as explained in Table 4.5, is in fact a backward selection, starting from the least restrictive structure (UN). In general, this procedure can be summarised as follows.

1. Order the variance-covariance structures according to the number of parameters in a decreasing order. (See Tables 4.5, 4.10 and 4.12).

2. Calculate the corresponding *p*-value (one- or two-sided) for each possible comparison. Note, that the *p*-value can only be calculated using the loglikelihood ratio for the nested models. Consult Table 4.3 to determine whether two structures are nested or not.

3. For the first row, remove those structures (from columns) for which the *p*-value is not larger than the significance level (e.g. $\alpha = 0.05$).

4. If there are two non-nested variance-covariance structures S1 and S2, search for a variance-covariance structure S3 that is nested in both S1 and S2.

   a. If there is no such S3 or both S1 and S2 are significantly better than S3, compare S1 and S2 using BIC and/or AIC.

   b. If only S1 (or S2) is significantly better than S3, choose S1 (or S2).

   c. If S3 is significantly better than S1 and S2, choose S3.

5. Continue with the next row and follow the procedure of step 3 and step 4 until all variance-covariance structures are considered.

4.2.1.3 Step 3. Interpretation of the Parameter Estimates

Once the best fitting variance-covariance structure is selected, the fixed part of the model can be evaluated. The analysis of the Proximity study starts with the specification of the fixed part of the model according to Equation (4.1). For the sake of discussion, suppose we choose the RI/RS model. The estimates of the regression parameters for the RI/RS model are displayed in Table 4.6.

Caution is advised when testing interaction terms that consist of a combination of dummy variables as in the Proximity study. According to the overall F-test in Table 4.6, the interaction *Occasion × Sex* is not statistically significant (p-value = 0.168). However, if the interest is to compare the change in average of *Proximity* between each pair of time points or between successive time points for males and females separately, the overall F-test may also be conservative. Instead, direct comparisons between different time points (that are planned before the analysis) may be performed. Furthermore, Konietschke and Brunner (2013) argued that there may be situations that the overall null hypothesis, using the overall F-test, may not be rejected, while some of the

TABLE 4.6

Proximity Study: ANOVA Table and REML Estimates of the Regression Coefficients from the RI/RS Model (Reference Categories: *Sex* = 0 and *Occasion* = 3)

Source	Numerator df	Denominator df	F-value	p-value
Intercept	1	47.83	114.056	0.000
Occasion	3	64.91	1.092	0.359
Sex	1	49.62	0.000	0.998
Occasion × Sex	3	69.60	1.732	0.168

Dependent Variable: *Proximity*.

Parameter	Estimate	Std. Error	df	t-value	p-value	95% Confidence Interval Lower Bound	95% Confidence Interval Upper Bound
Intercept	0.62	0.07	51.68	9.38	0.000	0.49	0.76
Occ_0	0.00	0.09	40.72	0.01	0.993	−0.18	0.18
Occ_1	0.09	0.08	86.44	1.19	0.238	−0.06	0.25
Occ_2	0.01	0.07	102.92	0.16	0.877	−0.13	0.15
Sex	0.04	0.13	53.17	0.27	0.788	−0.23	0.31
$Occ_0 \times Sex$	−0.22	0.18	42.02	−1.25	0.217	−0.57	0.13
$Occ_1 \times Sex$	−0.01	0.16	97.22	−0.03	0.978	−0.32	0.31
$Occ_2 \times Sex$	0.08	0.14	102.92	0.59	0.559	−0.19	0.35

Dependent Variable: *Proximity*.

individual tests lead to significant results. The reverse might happen as well, that is a significant overall F-test may occur, while none of the individual tests lead to a significant result (see Gabriel, 1969; Hsu, 1996).

In principle, any pairwise comparison can be made from Table 4.6. However, not all corresponding p-values can directly be read from this table. Because male and the last occasion are the reference categories, Table 4.6 still allows comparing the average of *Proximity* at occasion 2 and occasion 3 among males. Hence, the estimated difference in average proximity scores $(\overline{Proximity}_m\,(Occ_2)$- $\overline{Proximity}_m(Occ_3))$ between occasion 2 and occasion 3 for males is $\hat{\beta}_3 = 0.01$ (the estimate of the regression coefficient corresponding to Occ_2), with p-value = 0.877. Unfortunately, this table does not provide p-values to compare the difference between occasion 0 and occasion 1 and between occasion 1 and occasion 2. By changing the reference occasion to occasion 1, it becomes possible to compare occasion 0 with occasion 1, and occasion 1 with occasion 2 and obtaining the p-values from the corresponding output.

Table 4.7 reports all six pairwise comparisons with the corresponding p-values. Note that these results are obtained after recoding the reference categories for both *Sex* and *Occasion*. For instance, the reader may check that the estimated difference $\overline{Proximity}_m\,(Occ_0)$- $\overline{Proximity}_m(Occ_1) = -0.09$ (p = 0.142) and $\overline{Proximity}_m\,(Occ_1)$- $\overline{Proximity}_m(Occ_2) = 0.08$ (p = 0.205).

To elaborate the calculation for females, let us define $Sexm = 0$ for females and $Sexm = 1$ for males. Also, we define Occ_1 as the reference category for *Occasion*. Table 4.8 shows the corresponding results, where the estimated difference $\overline{Proximity}_f\,(Occ_0)$- $\overline{Proximity}_f(Occ_1) = -0.31$ (p = 0.006) and $\overline{Proximity}_f$ (Occ_1)- $\overline{Proximity}_f(Occ_2) = 0.00$ (p = 0.974). By changing the reference category to occasion 3, the estimated difference in average proximity scores between occasion 2 and occasion 3 can be determined, which equals to $\overline{Proximity}_f$ (Occ_2)- $\overline{Proximity}_f(Occ_3) = 0.09$ (p = 0.442).

TABLE 4.7

Proximity Study: Estimated Differences in Average Proximity Score and their Corresponding p-values

Sex	Reference	Comparison $Occ_i - Occ_j$	Estimated Difference (p-value*)
Male	Occasion 1	$i = 0, j = 1$	-0.09 ($p = 0.142$)
		$i = 1, j = 2$	0.08 ($p = 0.205$)
	Occasion 3	$i = 2, j = 3$	0.01 ($p = 0.857$)
Female	Occasion 1	$i = 0, j = 1$	-0.31 ($p = 0.006$)
		$i = 1, j = 2$	0.00 ($p = 0.974$)
	Occasion 3	$i = 2, j = 3$	0.09 ($p = 0.442$)

* Uncorrected p-value between brackets for significance level $\alpha = 0.05$.

TABLE 4.8

Proximity Study: REML Estimates of the Regression Coefficients from the RI/RS Model (Reference Categories: Female (*Sexm* = 0) and First Occasion (*Occasion* = 1))

Parameter	Estimate	Std. Error	df	*t*-value	*p*-value	95% Confidence Interval Lower Bound	Upper Bound
Intercept	0.75	0.13	86.28	5.58	0.000	0.48	1.01
Occ_0	−0.31	0.11	77.51	−2.81	0.006	−0.53	−0.09
Occ_2	0.00	0.12	82.69	0.03	0.974	−0.23	0.24
Occ_3	−0.09	0.14	100.18	−0.63	0.532	−0.36	0.19
Sexm	−0.03	0.15	82.79	−0.20	0.841	−0.34	0.28
$Occ_0 \times Sexm$	0.21	0.13	79.29	1.72	0.090	−0.03	0.46
$Occ_2 \times Sexm$	−0.09	0.13	82.63	−0.64	0.525	−0.35	0.18
$Occ_3 \times Sexm$	−0.00	0.16	97.22	−0.03	0.978	−0.32	0.32

Dependent Variable: *Proximity*.

Recall that six pairwise comparisons need to be tested. To control for the overall type I error rate, a multiple comparison procedure correction must be performed. We briefly review some multiple comparison procedures that are relatively easy to carry out. For an elaborated comparison between these methods see, e.g., Afshartous and Wolf (2007), Liu et al. (2010), and Sauder and DeMars (2019).

Bonferroni Correction

Bonferroni correction is achieved by dividing the overall significance level by the number of k performed tests. The idea is to provide a more conservative test. If there are k independent tests, the significance level is divided by k so that each null hypothesis is tested with a new (corrected) significance level $\alpha^* = \alpha/k$.

Sidak-Bonferroni Correction

This correction method is another approach that controls the familywise error rate. The (corrected) level of significance is calculated by $\alpha^* = 1 - \sqrt[k]{1-\alpha}$, where α is the overall level of significance and k is the number of tests. This method produces a less conservative result compared with the Bonferroni method, because the adjusted level of significance is slightly higher.

Holm-Bonferroni correction

The Holm correction method is also an adjusted version of the Bonferroni method to account for the conservative nature of the Bonferroni procedure.

The procedure leads to the probability of type I errors for at least one test that is equal to the overall significance level α. The Holm can be summarised as follows:

Order the p-values of the family of hypotheses in an ascending order. For k tests, as an example, the p-values can be written as $p_1 \leq p_2 \leq \ldots \leq p_k$.

a. Define the adjusted significance level

$$\alpha_1^* = \frac{\alpha}{k}, \alpha_2^* = \frac{\alpha}{k-1}, \alpha_3^* = \frac{\alpha}{k-2} \ldots \alpha_k^* = \alpha.$$

b. For every $i = 1, \ldots, k$, if $p_i \leq \alpha_i^*$, the null hypothesis for the ith test is rejected in favour of the alternative hypothesis; Otherwise, the null hypothesis is not rejected for that i^{th} test and all successive tests.

It is very important to realise that any pairwise comparisons should be planned prior to analysis to prevent data dredging. The easiest way is to use the Bonferroni adjustment by dividing the significance level (e.g. $\alpha = 0.05$) by the number of tests (six tests in the Proximity study), that is the corrected significance level becomes $\alpha^* = 0.05/6 = 0.008$. After applying the correction, the only significant change (at $\alpha^* = 0.008$) in estimated average proximity score is between occasion 0 and occasion 1 for females.

In the Proximity study, the main effects *Sex* and *Occasion* should be kept in the model irrespective of being significant or not, because the interest is the effect over time for male and female teachers separately. The only term that is worthwhile to test is therefore the interaction term. However, as we have noticed previously, care should be taken before removing interaction terms as demonstrated with the analysis of this example. From Table 4.6, it looks as if all interaction terms are not significant at 0.05 level, but after closer inspection, it appears that there is a significant increase in average proximity score from occasion 0 to occasion 1 among females.

As a final note, it is recommended to use the maximum likelihood estimates and the corresponding -2LL rather than the restricted -2LL to compare models that differ in their fixed parts (Verbeke and Molenberghs, 2000, Snijders and Bosker, 2012). For large samples, however, both the ML and REML approaches give unbiased estimates of the fixed parameters. If the final model is obtained, all parameters should be estimated using the REML approach as the very last step.

4.2.2 Best Practice for the Analysis of the Growth Study

In the Growth study as specified in Equation 3.1, which is specified as

$$Distance_{ij} = \beta_0 + \beta_1\,Sex_i + \beta_2\,Age_{ij} + \beta_3\,Sex_i \times Age_{ij} + V_{ij},$$

the observed correlation-matrix is

$$\mathbf{V}_{corr-observed} = \begin{pmatrix} 1 & 0.63 & 0.71 & 0.60 \\ 0.63 & 1 & 0.64 & 0.76 \\ 0.71 & 0.64 & 1 & 0.80 \\ 0.60 & 0.76 & 0.80 & 1 \end{pmatrix} \tag{4.2}$$

Recall that this matrix can be obtained by first fitting the standard OLS to Equation 3.1 and then calculating the pairwise correlation of residuals across time points. From Equation 4.2, it is seen that the residuals between time points are correlated. The scatter plot matrix in Figure 4.2 visualises the bivariate relationship between residuals across time points.

From both Figure 4.2 and the correlation matrix, it can be argued that the RI model is presumably sufficient to explain the correlation structure of the responses because the observed correlations are much alike. We remind the readers that the RI model assumes the same correlation between two time points. We can also consider a more complex correlation structure such as the RI/RS model that does not necessarily impose identical correlation structure across time. In the following, we will consider a number of alternative covariance structures for the Growth study.

Following the marginal approach, an obvious way to estimate the correlation structure is to impose no restrictions on the variances and covariances (and so correlations), that is the V-matrix is left unstructured (i.e. UN variance-covariance structure). The estimated V-matrix in this case is

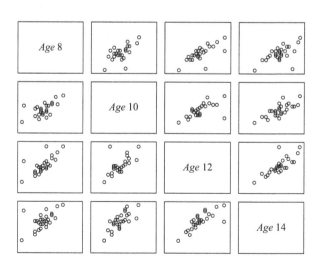

FIGURE 4.2
Scatter plot matrix of responses in the Growth study.

$$\mathbf{V}_{corr-UN} = \begin{pmatrix} 1 & 0.57 & 0.66 & 0.52 \\ 0.57 & 1 & 0.58 & 0.72 \\ 0.66 & 0.58 & 1 & 0.73 \\ 0.52 & 0.72 & 0.73 & 1 \end{pmatrix}, \tag{4.3}$$

where each of its elements is estimated as a separate parameter. Other variance-covariance structures such as AR (1) or Toeplitz could be considered as well. The main reason relates to the rather large number of parameters that need to be estimated when an unstructured V-matrix is specified. In the Growth study, for example, 10 variances and covariances should be estimated. Together with the regression parameters in the fixed part, a total of 14 parameters are estimated. In contrast, six parameters are only estimated (four regression parameters plus ρ and the first-level variance σ^2) if an AR (1) with homogeneous variances is used. The estimated V-matrix for this structure is

$$\mathbf{V}_{corr-AR(1)} = \begin{pmatrix} 1 & 0.62 & 0.39 & 0.24 \\ 0.62 & 1 & 0.62 & 0.39 \\ 0.39 & 0.62 & 1 & 0.62 \\ 0.24 & 0.39 & 0.62 & 1 \end{pmatrix}, \tag{4.4}$$

The Toeplitz variance-covariance structure (either homogeneous or heterogeneous variances) is always an interesting candidate because it is very close to unstructured but requires less parameters. For the Growth study, the estimated correlation V-matrix (for Toeplitz homogeneous) is

$$\mathbf{V}_{corr-toeplitz} = \begin{pmatrix} 1 & 0.64 & 0.70 & 0.48 \\ 0.64 & 1 & 0.64 & 0.70 \\ 0.70 & 0.64 & 1 & 0.64 \\ 0.48 & 0.70 & 0.64 & 1 \end{pmatrix} \tag{4.5}$$

which has four parameters (one variance and three covariances). Comparing the estimated correlation matrix of Equations (4.3)–(4.5) with the observed correlation matrix of Equation 4.2, one can see that the AR (1) shows the largest deviation from the observed correlation matrix. In general, however, it would be difficult to determine the best correlation matrix without a proper goodness of fit test.

Table 4.9 summarises the (restricted) -2LL values of the candidate variance-covariance structures of the Growth study together with the BIC and AIC. A comparison is made between models with alternative V matrices in Table 4.10.

TABLE 4.9

Growth Study: Candidate Variance-Covariance Structures, Ordered According to the Number of Parameters

	Variance-Covariance Structure	Par	-2LL (REML)	BIC	AIC
1	UN	14	424.55	470.99	444.55
2	Random intercept + slope	8	432.58	451.16	440.58
3	Toeplitz	8	429.39	447.97	437.39
4	RI/CS	6	433.76	443.05	437.76
5	AR (1) homogeneous	6	444.59	453.88	448.59
6	Independent (constant variances)	5	483.56	488.20	485.56

TABLE 4.10

Growth Study: the Scheme to Determine the Best Variance-Covariance Structure

	Model	Par	1	2	3	4	5	6
1	UN	14	X	Df=6 Chi2=8.04 p=0.24	Df=6 Chi2=4.85 p=0.56	Df=8 Chi2=9.21 p=0.32	Df=6 Chi2= 20.04 p=0.01	Df=9 Chi2=59.01 p=0.00
2	RI/RS	8		X	NP	Df=2 Chi2=1.18 p=0.28	NP	Df=3 Chi2=50.98 p=0.00
3	Toep	8			X	Df=2 Chi2=4.37 p=0.11	Df=2 Chi2=15.20 p=0.00	Df=3 Chi2=54.17 p=0.00
4	RI/CS	6				X	NP	Df=1 Chi2=49.80 p=0.00
5	AR (1)	6					X	Df=1 Chi2=38.97 p=0.00
6	Ind	5						X

NP, not possible with likelihood ratio.

From this Table, it becomes apparent that the UN variance-covariance structure is not significantly better than the RI (or CS), RI/RS and Toeplitz (first row). Furthermore, the RI/RS and Toeplitz are not hierarchically related, but they are not significantly superior to the RI (or CS), with p-value 0.28 and 0.11, respectively. Following the procedure described in Section 4.2.1, it can hence be concluded that the RI model is the most parsimonious model that fits the data equally well as the UN model. Also, note that the BIC-value of the RI model is smallest, while the AIC of the Toeplitz homogeneous is smallest.

The estimation and interpretation of the fixed part of the final model (i.e. the RI model) have already been discussed in Section 3.2. The conclusion of the study is that girls and boys significantly differ from each other in their growth, because the interaction between *Sex* and *Age* is significant at $\alpha = 0.05$ level ($p = 0.014$).

In some statistical packages (like SPSS), a random-effects specification may lead to report zero estimates of one or more random-effects variances. Other statistical packages present negative variances. In the next section, we elaborate on this issue by considering the Salsolinol study that is also analysed by Landau and Everitt (2004).

4.2.3 Best Practice for the Analysis of the Salsolinol Study

In the Salsolinol study, two groups of subjects, one with moderate and the other with severe dependence on alcohol, had their Salsolinol secretion levels measured (in log (mmol)) on four consecutive days (Landau and Everitt, 2004). Primary interest is whether the groups evolved differently over time. Thus, an interaction term must be included in the model. In this study, there were no missing observations, and so, the design is balanced in time with equidistant time points. Figures 4.3a and b show the subject-specific and average profiles, respectively. The former displays a considerable within and between-subject variation, indicating that the data are correlated, while the latter suggests specifying the following model that includes an interaction between *Time* and *Group*.

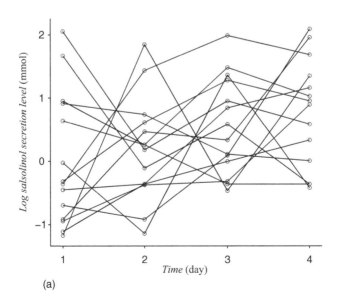

(a)

FIGURE 4.3
(a) Subject-specific profiles for the Salsolinol study. *(Continued)*

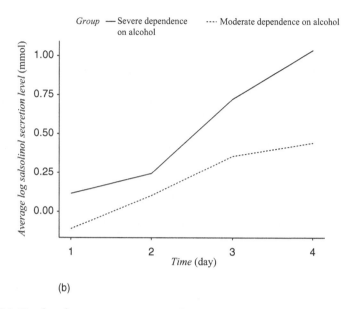

(b)

FIGURE 4.3 (Continued)
(b) Average profiles for moderately and severely dependent group in the Salsolinol study.

$$log\left(Salsolinol\right)_{ij} = \beta_0 + \beta_1\,Group_i + \beta_2\,Time_{ij} + \beta_3\,Group_i \times Time_{ij} + V_{ij}.$$

Before selection of any variance-covariance structure, let us look at the observed correlation between responses (equation 4.6), where we notice that some correlations are negative.

$$\mathbf{V}_{corr-observed} = \begin{pmatrix} 1 & -0.06 & 0.16 & -0.72 \\ -0.06 & 1 & 0.00 & 0.25 \\ 0.16 & 0.00 & 1 & 0.00 \\ -0.72 & 0.25 & 0.00 & 1 \end{pmatrix} \tag{4.6}$$

If we start with an RI model (as the simplest choice), various statistical packages including SPSS will not be able to estimate the random-intercept variance properly, because the programme does not converge so that the results cannot be trusted (although some others may still produce a negative variance, which is also not a proper estimate). The reason is that the correlation between responses in any RI model must always be positive. Recall that, in an RI model, the correlation between responses is equal to the ICC (see equation 3.9), which is always positive. The observed correlations in Equation (4.6), however, show some negative values; for example, the observed correlation between time point 1 and time point 4 is strongly negative (-0.72). Hence, the RI model is certainly not an appropriate choice for the Salsolinol study.

The RI/RS model can be fitted to the data as well and further compared with other alternative variance-covariance structures. Table 4.11 summarizes the (restricted) -2LL values of the candidate variance-covariance structures of the Salsolinol study together with the BIC and AIC. A comparison between the RI/RS and an OLS leads to a nonsignificant result (Table 4.12: *p*-value = 0.18), suggesting superiority of the OLS over the RI/RS model.

TABLE 4.11

Salsolinol Study: Alternative Variance-Covariance Structures, Ordered According to the Number of Parameters

	Model	nr. Of Estimated Parameters	-2LL (REML) BIC, AIC Respectively, Between Brackets
1	UN	14	131.89 (171.40, 151.89)
2	Toeplitz(heterogeneous)	11	131.93 (159.59, 145.93)
3	Compound symmetry (heterogeneous)	9	143.72 (163.47, 153.72)
4	Toeplitz (homogeneous)	8	135.25 (151.06, 143.25)
5	Random intercept + slope	8	140.99 (156.79, 148.99)
6	Compound symmetry (homogeneous)	6	145.56 (153.46, 149.56)
7	Independence (constant variances)	5	145.84 (149.79, 147.84)
8	Random intercept	not possible	

p-values based on likelihood ratio.

TABLE 4.12

Salsolinol Study: the Scheme to Determine the Best Variance-Covariance Structure

	Model	Par	1	2	3	4	5	6	7
1	UN	14	X	0.998	0.04	0.76	0.17	0.09	0.12
2	Toeplitz(heterogeneous)	11		X	0.003	0.34	NP	0.02	0.00
3	Compound symmetry (heterogeneous)	9			X	NP	NP	0.61	0.71
4	Toeplitz (homogeneous)	8				X	NP	0.01	0.01
5	Random intercept + slope	8					X	NP	0.18
6	Compound symmetry (homogeneous)	6						X	1
7	Independence (constant variances)	5							X

Each cell contains a *p*-value (compare with table 4.10).
NP, not possible with likelihood ratio.

Moreover, a comparison between the OLS and UN structure leads to a non-significant result (Table 4.12: p-value = 0.12). Consequently, the OLS would incorrectly be chosen as the most appropriate model if one only relies on the random-effects models (i.e. subject-specific approach).

From Table 4.12, it becomes clear that the Toeplitz with homogeneous variances is the most appropriate covariance structure. The reader may follow the model selection procedure described above. On the one hand, this model, however, does not have the lowest BIC demonstrating that the BIC sometimes incorrectly favours the simplest model. On the other hand, the AIC has the smallest value so that it leads to the same conclusion as the likelihood ratio test. This is, nevertheless, not always the case.

Choosing the homogeneous Toeplitz as the correct variance-covariance structure and performing a backward procedure leads to a marginal model without interaction and a significant group effect at 10% level (difference of 0.39 mmol on a logarithmic scale with p-value = 0.069). This result would not be obtained had the OLS been used (difference of 0.33 mmol on a logarithmic scale with p-value = 0.157).

The estimated correlation matrix between responses ($\mathbf{V}_{corr}$) is

$$\mathbf{V}_{corr-estimated} = \begin{pmatrix} 1 & -0.03 & 0.20 & -0.62 \\ -0.03 & 1 & 0.03 & 0.20 \\ 0.20 & 0.03 & 1 & 0.03 \\ -0.62 & 0.20 & 0.03 & 1 \end{pmatrix}, \tag{4.7}$$

which is in accordance with the observed one (Equation 4.6).

The abovementioned example clearly shows that a random-effects model might not always be sufficient to properly analyse longitudinal data. To choose the most suitable model for correlated data, comparisons should be made between random-effects models (with/without serial correlations) and marginal models (with alternative variance-covariance structures).

4.3 Multiple Imputation of Missing Observations: Building Imputation Models

In the previous chapter, we discussed various methods of handling missing data without considering any imputation technique. This section instead focuses on multiple imputation (MI) of missing observations in longitudinal studies. MI is particularly useful when baseline or time-varying independent variables have missing observations. Although the missingness of dependent variable does not necessarily require any imputation action, MI is still appealing for conducting sensitivity analysis to investigate the robustness of the results to the MAR assumption. Additionally, imputation of missing observations in the dependent variables can have a direct effect on statistical power as compared to the direct likelihood approach. As discussed in the previous chapter,

→ *if there are missing observations in both the dependent and time-varying independent variables, multiple imputation also becomes an attractive alternative to direct likelihood approach for handling missing data issues.*

Several imputation routines are directly accessible from many statistical packages, but these do not necessarily produce good imputations for missing observations. If, for example, the imputed values are incorrect, an analysis based on the imputed data will produce biased estimates (e.g. the treatment effect estimate), which, in turn, invalidate the results. Therefore, a big effort is required to obtain good imputations.

Obtaining good imputations can be translated into imputation models that are (nearly) correctly specified. Therefore, incorrect imputation models are the main cause for distortion in the analysis.

→ *As a general rule, the data structure should be reflected in the imputation model.*

For example, if the variance-covariance structure assumes dependency between time points, for example AR (1), or there exist potential confounders or modifiers, the imputation model should take these facts into account when missing observations are imputed. We therefore build imputation models based on the *inclusive strategy* that allows the imputation model being more general than the analysis model; that is, the analysis model is more restrictive than the imputation model. Figure 4.4 depicts the situation where the inner oval includes the *true* analysis model. Here, we assume that the analysis model could be chosen from the inner oval if there were no missing observations at all. The imputation model is thus chosen from an extended class of models such that the class of analysis model is its subset.

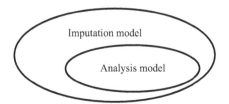

FIGURE 4.4
Imputation and analysis model based on the inclusive strategy.

To elaborate the general idea, consider the Growth study wherein children were measured at four time points. For the variance-covariance structure of the analysis model in this study, a choice should be made that varies from unstructured (i.e. the least restrictive) to independence (i.e. the most restrictive). If, on the one hand, the imputation model assumes an unstructured variance-covariance structure, any variance-covariance structure for the analysis model (e.g. the compound symmetry) is *at most* as complex as the variance-covariance structure for the imputation model, and so, the inclusive strategy is not violated. If, on the other hand, the compound symmetry structure is used for the imputation, but the analysis model uses an unstructured variance-covariance structure, the imputation model will be more restrictive than the analysis model and so the inclusive strategy is violated. Another example is the presence of auxiliary variables such as secondary dependent variables or proxies. Inclusion of these variables into the imputation model is not harmful in general, as these variables are not necessarily a part of the main analysis. Construction of such imputation models is based on the inclusive strategy because the imputation model makes less assumptions than the analysis model (i.e. the former has more parameters than the latter). Nevertheless, this does not mean that it is always possible or practically feasible to construct a very general imputation model simply because of various limitations in the data at hand. For instance, a very large imputation model with many parameters can be over-parameterised so that these parameters cannot be estimated anymore. Another example is when the auxiliary variable has a high rate of missingness making the imputation process cumbersome and less precise.

In the subsequent sections, we first distinguish between the so-called wide data format and long data format and then elaborate on how to construct imputation models for imputing missing observations using either format. We will also provide some recommendations on the construction of imputation models in general.

4.3.1 Data Format: Wide versus Long

Any longitudinal dataset can be represented in two formats. When the data are arranged in the wide format, each subject occupies only one row in the data matrix. The dependent variable at each time point is thus reported in a separate column of the matrix (the same rule applies to any time-varying independent variable as well). Table 4.13 shows the records of two cases in the Proximity study.

Here, the proximity scores at baseline and the other three occasions are reported as separate variables in separate columns (i.e. Occ_0, Occ_1, Occ_2 and Occ_3). Note that the time variable is implicit in the number of repeated measures and will not be explicitly provided in this format. *Sex* as a baseline or time-invariant independent variable is reported in a separate column.

Alternatively, each subject occupies more than one row when the data are arranged in the long format. Each row is one time point per subject and each subject will have data in a block of rows. Baseline variables, such as *Sex*, will have the same value in all these rows (see Table 4.14 where the above two cases are rearranged in the long format).

In the Proximity study, scores were measured at four occasions so that a block of four rows corresponds to each subject (as opposed to the wide format where one row corresponds to one subject). Here, for each subject, the proximity score (i.e. the dependent variable) is displayed consecutively in the last column of Table 4.14. To distinguish the dependent variable (and any other time-varying independent variable) for each subject per occasion, two additional variables must be defined in the dataset. In Table 4.14, variables '*Teacher ID*' and '*Time*' represent the case number and occasion, respectively. Finally, the baseline variable '*Sex*' is repeated per block of rows (four times in this study).

The imputation task can be performed in either layout, but an analysis based on the random-effects models requires the long format of the data. Most statistical packages have now built-in functions that allow going from one direction to the another (see the assignment for an example in SPSS and R).

TABLE 4.13

Proximity Study: A Schematic Representation of Two Cases in the Wide Format

			Proximity Score			
Row Number	Teacher ID	*Sex*	Occ_0	Occ_1	Occ_2	Occ_3
1	3	Male	.	1.10	0.97	1.06
2	7	Female	0.22	0.29	.	0.54

TABLE 4.14

Proximity Study: A Schematic Representation of Two Cases in the Long Format

Row number	*Teacher* ID	*Time* (Occasion)	Sex	*Proximity* score
1	3	0	Male	.
2	3	1	Male	1.10
3	3	2	Male	0.96
4	3	3	Male	1.06
5	7	0	Female	0.22
6	7	1	Female	0.29
7	7	2	Female	.
8	7	3	Female	0.54

4.3.2 Imputation in Wide Format: The Classical Approach

Multiple imputation in a wide format is the most common approach to impute missing observations in longitudinal studies. This is because (i) the imputation task is relatively straightforward, and (ii) it is widely available in mainstream statistical packages. For instance, the MI procedures in SPSS and SAS, by default, assume that the dataset is in the wide format when the imputation task is executed.

To explain the imputation procedure in the wide format, let us consider the Proximity study wherein the proximity score was measured at four occasions. Notice that the dependent variable is expressed in four variables (e.g., y_1, y_2, y_3 and y_4), each of which represents the proximity scores of an occasion. The dataset also contains the teacher's sex as an independent variable. Studying the pattern of missing data reveals a non-monotone pattern for the proximity scores (see Figure 3.9), so that an imputation model is required for every occasion.

Like Chapter 2, we use the MICE approach (i.e. the conditional imputation approach) for the imputation purpose.

→ *It should be noted that the imputation process is blind to the role of variables in the analysis model; that is, the MICE approach does not distinguish between the dependent and independent variables during imputation.*

Hence, the imputation procedure, explained below, can readily be extended when independent variables such as teacher's sex have missing observations (see also Chapter 2).

Let us define the following four imputation models for the proximity score at different occasions:

$$
\begin{aligned}
y_1 &= \alpha_{10} + \alpha_{11} \times Sex + \alpha_{12} \times y_2 + \alpha_{13} \times y_3 + \alpha_{14} \times y_4 + \epsilon_1 \\
y_2 &= \alpha_{20} + \alpha_{21} \times y_1 + \alpha_{22} \times Sex + \alpha_{23} \times y_3 + \alpha_{24} \times y_4 + \epsilon_2 \\
y_3 &= \alpha_{30} + \alpha_{31} \times y_1 + \alpha_{32} \times y_2 + \alpha_{23} \times Sex + \alpha_{34} \times y_4 + \epsilon_3 \\
y_4 &= \alpha_{40} + \alpha_{41} \times y_1 + \alpha_{42} \times y_2 + \alpha_{43} \times y_3 + \alpha_{44} \times Sex + \epsilon_4
\end{aligned}
\tag{4.8}
$$

where α's are regression weights, and each error term ϵ_i, $i = 1, \ldots, 4$ follows a normal distribution with mean zero and variance σ_i^2, $i = 1, \ldots, 4$. Here, as an example, the missing proximity scores at the first occasion (i.e. at baseline denoted by y_1) are imputed using the proximity scores at the second, third and fourth occasions together with teacher's sex. The latter variable is added to each imputation model because the main goal of the study is to compare the proximity score of successive occasions within male and female teachers. If *Sex* is excluded from the imputation Equations 4.8, it would implicitly assume that there is no difference, on average, between the proximity score of male and female teachers within each occasion. When *Sex* is then included in the analysis model (after imputation), the imputation models will be more restrictive than the analysis model, which is a violation of the inclusive strategy.

The abovementioned imputation models also imply an unstructured variance-covariance structure for the repeated measurements in the imputation process. Because the proximity score at each occasion is assumed to follow a normal distribution conditional on the proximity score at other occasions, the joint distribution of proximity score (for all occasions) follows a multivariate normal distribution (see, Johnson and Wichern, 2014). It therefore can be argued that imputations are drawn from a multivariate normal model, which has an unstructured variance-covariance structure. It is essentially less restrictive than any other variance-covariance structure that may be used to analyse the imputed data (and so no violation of the inclusive strategy).

After defining the imputation models, the next step is to create imputations for missing proximity scores under Equation 4.8. Following the recommendation of White et al. (2011), the number of imputations should be close to the percentage of incomplete cases if the fraction of missing information (FMI, see Chapter 2, p. 105) is lower than the percentage of incomplete cases.

→ *We have thus created 70 imputed datasets as only 30 percent of subjects had the proximity scores in all four occasions and the estimated FMI was below 0.70 for all incomplete variables.*

A very important step after creating imputations is the diagnostic monitoring of the convergence of the algorithm and investigation of the plausibility of the imputations. The former indicates potential issues with the imputation process, while the latter evaluates inconsistent imputations.

If the number of iterations is insufficient or if convergence is not achieved, the imputations are incorrect in general. The convergence of the MICE algorithm can be checked, for example, by plotting separately the mean and standard deviation of the imputed values across the iteration cycle for each incomplete variable. When there is no pattern across different imputations, it can be concluded that the algorithm has converged. As an example, the convergence of the algorithm for the imputed proximity scores at occasion 3 is visualised in Figure 4.5. Different lines within each graph represent the mean (top) and standard deviation (bottom) of the imputed values for each imputed set. The figure displays stable imputations without any trend after the first five iterations confirming the convergence of the algorithm. Examples of non-convergence within the MICE algorithm are provided in the study by van Buuren (2018, pp. 187–189). In practice, about 5–20 iterations are usually adequate for convergence.

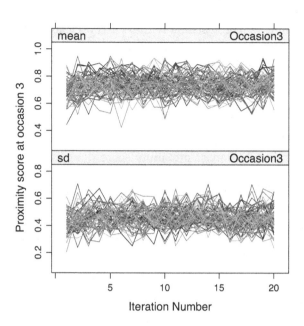

FIGURE 4.5
Mean (above) and standard deviation (bottom) of the imputed proximity scores at occasion 3 against the iteration number for 70 imputations in the Proximity study.

Implausible imputations imply that generated values are inconsistent with the data at hand or even impossible, such as pregnant fathers! It is therefore important to assess whether the generated imputations are plausible. One way is to perform the standard diagnostic checks on each imputed data set comparing the observed and imputed values. In the Proximity study, for instance, the scatter plot matrix of residuals (like Figure 4.1) can be used to assess the possible discrepancy between the observed and imputed data. No indication of a correlation between time points across all imputed data sets suggest a problem with the imputation models, as the proximity scores are supposed to be correlated. Figure 4.6 displays the residual plots for a randomly chosen imputed data set after applying the standard OLS procedure to Equation 3.12. Similar to the observed data (Figure 4.1), the imputed data in Figure 4.6 show a positive correlation between time points though the correlations decrease as the time span increases between two occasions. Van Buuren (2018, chapter 6.6) presents a nice overview of diagnostics for imputed data within the MICE framework.

After completing the diagnostic checks, each imputed data set is analysed as if it was complete, the results of which are then pooled to form a single inference. For example, a random-effects model can be fitted to each imputed data set, and the corresponding estimates (i.e. regression coefficients) can subsequently be combined. Using the main-stream statistical

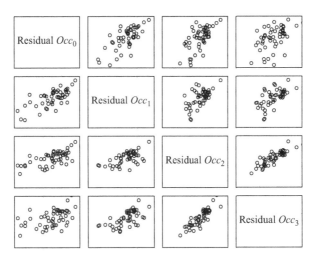

FIGURE 4.6
Scatter plot matrix of residuals from a randomly chosen imputed data set in the Proximity study.

packages/software such as SPSS or SAS, it is however an uneasy task to select among different models, for example to select between a model with an unstructured and a compound symmetry variance-covariance structure or between a model with and without interaction. The main reason is that most statistical packages do not offer the multivariate pooling procedures. As a result, comparing different random-effects models using the likelihood ratio test is currently lacking in the main-stream statistical packages/ software. It must be mentioned that the R package *mice* provides several multivariate pooling procedures known as D_1, D_2 and D_3 approaches for model building in the presence of multiple imputations. However, their current implementation is limited to uncorrelated data. We expect that new procedures are available soon for correlated data, as this is an ongoing research area.

For the Proximity study, we thus use model 3.12 that was developed in Section 4.2.1 for each imputed data set. Table 4.15 shows the pooled results from 70 imputations in the proximity data. Note that the imputed data sets are in the wide format, so they must be restructured to the long format before applying the random-effects model. As expected, the results are comparable to the results from the fitted random-effects model in Section 4.2.1 (see Table 4.6) because missing observations are limited to the dependent variable only.

Multiple imputation of missing observations in a wide format easily accommodates the presence of missing data in the baseline or time-varying independent variables. As an example, suppose there were some missing observations in the baseline variable *Sex*. An additional imputation model

TABLE 4.15

Proximity Study: Estimate of the Regression Coefficients from Multiple Imputation (70 Imputations) in the Wide Format

					95% Confidence Interval			
Parameter	Estimate	Std. Error	t-value	p-value	Lower Bound	Upper Bound	Fraction Missing Info.	Relative Increase Variance
Intercept	0.65	0.07	8.93	0.000	0.51	0.79	0.20	0.25
Occ_0	−0.02	0.09	−0.25	0.802	−0.19	0.15	0.20	0.25
Occ_1	0.06	0.08	0.78	0.438	−0.09	0.21	0.31	0.44
Occ_2	−0.03	0.07	−0.44	0.662	−0.16	0.10	0.36	0.55
Gender	0.03	0.15	0.23	0.816	−0.26	0.32	0.30	0.42
$Occ_0 \times$ Gender	−0.22	0.18	−1.19	0.236	−0.57	0.14	0.34	0.50
$Occ_1 \times$ Gender	0.01	0.17	0.04	0.972	−0.32	0.34	0.46	0.85
$Occ_2 \times$ Gender	0.09	0.14	0.63	0.533	−0.19	0.37	0.45	0.81

joins the set of imputation Equations 4.8 such that *Sex* is imputed from the proximity scores of all occasions (i.e. y_1, y_2, y_3, y_4). As an example, the following logistic regression model can be defined for the purpose of imputing *Sex*

$$log\left\{\frac{Pr\left(Sex = female\right)}{Pr\left(Sex = male\right)}\right\} = \alpha_0 + \alpha_1 y_1 + \alpha_2 y_2 + \alpha_3 y_3 + \alpha_4 y_4$$

Time-varying independent variables with missing observations are also treated in the same manner, that is an additional imputation model is defined for each time-varying independent variable with missing observations.

Although imputation of repeated measurements in the wide format is rather simple, it is not always possible or practically feasible to consider a saturated multivariate model (with an unstructured covariance matrix) for the imputation task. For instance, when the number of repeated measurements becomes large (e.g. 20 time points), the conditional imputation models implicitly imply a 20-variate normal model with a 20 × 20 covariance matrix requiring estimation of 210 variance and covariance parameters (i.e. 20 variances plus 190 covariances). This might lead to overparameterisation of the model, particularly when the number of subjects is not large. Beerens et al. (2018) conducted a study to compare the large and small-scale care facilities when the study included 81 repeated measurements of elderly people. This study suffered from missing observations, but it was not feasible to generate imputations from the wide format procedure. Tan, Jolani, and Verbeek (2018) developed a practical solution by imposing some constrains to the covariance structure and generated imputations for the missing observations in this study.

In conclusion, we highlight that a balance should be kept between an overly simplistic model and a complex model (e.g. a model with many high order interactions) when the imputation task is performed. Moreover, additional information (e.g. secondary dependent variables) may be used to enrich the imputation model.

4.3.3 Imputation in a Long Format: The Multilevel Approach

In addition to the wide format, missing observations in longitudinal studies can also be imputed when the data are in the long format. Here, random-effects models are used to impute missing observations. For instance, the following imputation model can be defined to impute missing observations of the proximity scores:

$$
\begin{aligned}
Proximity_{ij} = {} & \beta_{0i} + \beta_1 Occ_{1ij} + \beta_2 Occ_{2ij} + \beta_3 Occ_{3ij} + \beta_4 Sex_i + \beta_5 Occ_{1ij} \\
& \times Sex_i + \beta_5 Occ_{2ij} \times Sex_i + \beta_5 Occ_{3ij} \times Sex_i + R_{ij}
\end{aligned}
\tag{4.9}
$$

which is a random-intercept model. This model (Equation 4.9), however, is not the only choice for imputation of proximity scores. More complex multi-level imputation models such as a random-intercept/random-slope (RI/RS) model could also potentially be considered.

Imputation of missing observations in the long format has several advantages. First, it is not needed to transpose data to the wide format for the purpose of imputation and then transform back to the long format for the analysis. Therefore, the long format of data is used for both imputation and analysis purposes, which is appealing in practice. Another advantage of imputation in the long format relates to the transparency about the inclusive strategy. Given an analysis model, it is immediately obvious whether the inclusive strategy is violated or not. Suppose the researcher uses a random-intercept model to analyse the data. The imputation model thus should be at least as complex as the random-intercept model. This implies that either a random-intercept or an RI/RS model can be specified for the imputation model, but models simpler than a random-intercept model violate the inclusive strategy and hence are inappropriate.

Although imputation based on the long format is attractive, various limitations are recognised. Foremost, multilevel imputation models are under-developed and still unavailable in the mainstream statistical packages such as SPSS and SAS. As an example, it is yet unclear how missing observations should be imputed using a random-effects model that allows for serial correlation. Moreover, theoretical properties of multilevel imputation models are not well understood (Resche-Rigon and White, 2018). Nevertheless, considerable progress has been made to fill some of the existing gaps by providing various routines in open-source statistical software such as R (see, e.g., Jolani et al., 2015; Audigier et al., 2018; Speidel, Drechsler and Jolani, 2020).

To briefly demonstrate the imputation procedure in the long format, we use the R packages *mice* and *lme4* to impute missing observations in the Proximity study. Following the arguments provided in Section 4.2.1, suppose the analysis is based on an RI/RS model that includes *Occasion*, *Sex* and their interaction as a fixed part in the analysis model. To impute missing observations in the proximity scores, the same RI/RS model is thus defined as the imputation model to fulfil the inclusive strategy. Here, inclusion of *Sex* and its interaction with *Occasion* (in the imputation model) preserve the association between proximity scores and teacher's sex when missing observations are imputed. Auxiliary variables such as secondary dependent variables, if available, could potentially be included in the imputation model as well to enrich the imputation phase.

Using the *2l.lmer* function in the R package *mice* (see, Jolani, 2018), missing observations in the proximity score were first imputed 70 times, and then, the analysis model (i.e. the RI/RS model) is fitted to each imputed data set using the *lmer* function in the R package *lme4*. Finally, all results are combined using the Rubin's rule. Code for accomplishing this in R is presented in the accompanying website.

The pooled results are presented in Table 4.16, which are similar to the results reported in Tables 4.6 and 4.15. It therefore can be concluded that when missing observations are limited to the dependent variable (i.e. the proximity score), different approaches, that is multiple imputation in the long format (Table 4.16), multiple imputation in the wide format (Table 4.15) and direct likelihood without imputation (Table 4.6) provide comparable estimates of regression coefficients and standard errors. The random nature of the imputation process also explains minor differences between the point estimates.

It should be noted that these approaches are comparable as long as the imputation model is not mis-specified. For example, if the interaction between *Sex* and *Occasion* is excluded from the imputation model in the long format, the estimates of regression coefficients might be subject to bias.

Multiple imputation of missing observations in the long format can also be extended to missing independent variables (e.g. time-varying independent variables) in addition to missing dependent variables. To illustrate this point briefly, imagine the student's performance score was also available as a time-varying independent variable such that the analysis model is a random-intercept model taking the form

$$Proximity_{ij} = \beta_{0i} + \beta_1 Occ_{1ij} + \beta_2 Occ_{2ij} + \beta_3 Occ_{3ij} + \beta_4 Sex_i + \beta_5 Performance_{ij}$$
$$+ \beta_6 Occ_{1ij} \times Sex_i + \beta_7 Occ_{2ij} \times Sex_i + \beta_8 Occ_{3ij} \times Sex_i + R_{ij} \quad (4.10)$$

TABLE 4.16

Proximity Study: Estimates of the Regression Coefficients from Multiple Imputation (70 Imputations) in the Long Format

	Estimate	Std. Error	t- value	p-value	df	Relative Increase Variance	Fraction of Missing Information
(Intercept)	0.62	0.06	10.30	0.000	144.48	0.22	0.19
Occ_0	0.01	0.08	0.09	0.397	151.91	0.18	0.17
Occ_1	0.09	0.07	1.30	0.171	131.63	0.30	0.24
Occ_2	0.02	0.07	0.31	0.38	113.74	0.43	0.31
Sex	0.03	0.12	0.23	0.389	117.82	0.39	0.29
$Occ_0 \times Sex$	−0.21	0.16	−1.30	0.171	136.89	0.27	0.22
$Occ_1 \times Sex$	0.01	0.15	0.09	0.397	107.04	0.48	0.34
$Occ_2 \times Sex$	0.09	0.13	0.69	0.315	104.37	0.51	0.35

where β_5 is the regression coefficient corresponding to the performance score of students. In addition to the proximity score, suppose the performance score is missing for some students at different occasions too. Using the MICE principle, both scores can be imputed as follows: the proximity score is imputed using the above model that includes *Occasion, Sex, Occasion* × *Sex* and *Performance* as predictors in the imputation model. Likewise, the performance score is imputed from *Proximity, Occasion, Sex* (and possibly the interaction between *Occasion* and *Sex*) using the following random-intercept imputation model

$$\begin{aligned} Performance_{ij} = {} & \alpha_{0i} + \alpha_1 Occ_{1ij} + \alpha_2 Occ_{2ij} + \alpha_3 Occ_{3ij} + \alpha_4 Sex_i + \alpha_5 Proximity_{ij} \\ & + \alpha_6 Occ_{1ij} \times Sex_i + \alpha_7 Occ_{2ij} \times Sex_i + \alpha_8 Occ_{3ij} \times Sex_i + R_{ij} \end{aligned} \quad (4.11)$$

Finally, as we discussed in earlier chapters, the underlying assumption for validity of all methods discussed in this chapter is the MAR missing data mechanism. Because this assumption cannot be verified from the observed data, the results should be treated as a starting point for sensitivity analyses. In Chapters 5–7, we demonstrate how the sensitivity analysis can be performed for repeated measurement designs.

4.4 Summary

This section summarizes essential steps that can be followed to build models in longitudinal studies. If there are no missing observations, or at most they are limited to the dependent variable (under the plausibility of the MAR assumption):

1. *Perform the exploratory analysis*: Individual and mean profile plots are produced (e.g. Figure 3.7 for the Proximity study; Figure 3.2 and 3.3 for the Growth study and Figure 4.3a, 4.3b for the Salsolinol study).
2. *Specify full relevant fixed part*: This is the fixed (systematic) part of the marginal or random-effects regression model. Relevant independent variables are selected based on substantive arguments and statistical consideration (e.g. whether an interaction is identified based on the mean profile plots).
3. *Fit standard OLS*: Regression parameters are estimated using the standard OLS approach (i.e. the OLS under the independence assumption), and the (correlated) residuals are kept.
4. *Obtain the observed correlation structure*: The $V_{corr\text{-}observed}$ is calculated (e.g. Table 4.2 for the Proximity study; Equation 4.2 for the Growth study and Equation 4.6 for the Salsolinol study)
5. *Compare several variance-covariance structures*: Several models are fitted (using REML) with varying variance-covariance structures, while the fixed (systematic) part of the model remains the same. Comparison is based on the likelihood ratio test for nested models (e.g. Table 4.5 for the Proximity study; Table 4.10 for the Growth study and Tables 4.11 and 4.12 for the Salsolinol study), but information criteria are used for non-nested models.
6. *Determine the best fixed part*: After choosing the most suitable variance-covariance structure, a selection procedure (e.g. backward) is used to choose the best fixed part of the model using the ML method. The final chosen model is refitted using the REML method (e.g. Table 4.6 for the Proximity study; Table 3.1 for the Growth study and see assignment 4.5.2 for the Salsolinol study).
7. *Interpret the regression parameter estimates*: The estimates of the final regression parameters are interpreted as indicated in Table 4.6 for the Proximity study and Table 3.2a and b for the Growth study.

If missing observations are extended to the independent variables, it is advisable to first perform the multiple imputation procedures of Sections 4.3.2 or 4.3.3 and then follow the above steps for the imputed data. Note that the very last results (i.e. the final estimates of the regression parameters) should be pooled to form a single result after multiple imputation. Furthermore, it should be acknowledged that the mainstream statistical packages (e.g. SPSS or SAS) are currently unable to pool the results at the model selection phase (steps 5 and 6), so that various variance-covariance structures cannot be compared with multiply imputed datasets in such packages. Nevertheless, open-source statistical software such as R opens new opportunities by providing routines to combine the results of multiple imputation in the model selection phase.

4.5 Assignments

4.5.1 Assignment

Consider the study about the orthodontic growth of boys and girls (Growth data, SPSS system file: Growthdata.sav).

Compare growth and growth velocity between boys and girls.

a. Run the OLS model (standard linear regression model with uncorrelated error terms)

$$Distance_{ij} = \beta_0 + \beta_1 Age_{ij} + \beta_2 Sex_i + \beta_3 Age_i \times Sex_i + V_{ij}$$

and plot the residuals against age. What can you say about the variance over time?

b. Produce a scatter-plot matrix of the residuals and determine the correlation matrix for the different time points. Do the correlations change with time? Which model will probably come out based on the exploratory analysis?

c. Consider some other (reasonable) alternative models with different covariance structures than the random intercept model and determine the most adequate model using the basic guidelines. Use other candidate variance-covariance structure than mentioned in Table 4.9.

d. Do the regression estimates depend on the candidate variance-covariance structure that you have considered?

e. Which model do you consider as best and what can be said about the comparison of *Distance* between boys and girls over time?

4.5.2 Assignment

Consider the Salsolinol study (SPSS system file: Salsolinol.sav). See Section 4.2.3 for a short description of the study.

a. Use the basic guidelines for model building to determine the best fitting model. Use other candidate variance-covariance structure than mentioned in Table 4.11.

b. Do the regression estimates depend on the candidate variance-covariance structure that you have considered?

c. Which model do you consider as best and what can be concluded about the comparison of salsolinol between the two dependency groups?

4.5.3 Assignment

Consider the Proximity study (SPSS system file: Teacher.sav). See for a short description Chapter 2.4.1. Suppose that we are interested in making a comparison between male and female teachers for each time point. Specifically, we are interested whether male and female teachers (do not) significantly differ in average proximity score over the whole period or only for part of the 4 years.

a. Use the basic guidelines for model building to determine the best fitting model. Use candidate variance-covariance structure other than that mentioned in Table 4.4.

b. Do the regression estimates depend on the candidate variance-covariance structure that you have considered?

c. Which model do you consider as best and what can be concluded about the comparison of proximity between the male and female teachers over time?

4.5.4 Assignment

Suppose that the researcher is also interested in the existence of serial correlation (probably on top of correlation induced by subject differences). Consider again the Proximity study. Take as a fixed part the full model with first-order interactions such as in Equation 3.1.

a. Fit a random intercept/random slope plus an AR (1) homogeneous in top of the RI/RS covariance structure. What can you about the parameter estimate $\hat{\rho}$? Discuss whether you can conclude that is (no) serial correlations.

b. Fit an AR (1) heterogeneous variance-covariance structure. What can you say about the significance of the parameter estimates? Discuss whether AR (1) is indicative for the existence of serial correlations.

c. Suppose that the RI/RS variance-covariance structure is the best fitting structure. What can you conclude about serial correlations?

4.5.5 Assignment

Consider the Life-event data (SPSS system file: Lifesubset.sav). (Nieboer et al., 1998). See Section 'short description of research and simulation study' and Chapter 6.1 for a short description of the study.

   a. Use the basic guidelines for model building to determine the best fitting model.
   b. Do the regression estimates depend on the candidate variance-covariance structure that you have considered?

4.5.6 Assignment

Consider the Proximity study (SPSS system file: Teacher.sav). See for a short description Chapter 2.4.1.

   a. Transpose the data to a wide format such that each teacher occupies only one row in the dataset. Save the data with a new name for the further use.
   b. Transpose the data that has been created in part a to the long format.

4.5.7 Assignment

Consider the Proximity study that is provided in a wide format (SPSS system file: Teacher_wide.sav). See 'short description of research and simulation study' and Chapter 2.4.1.
   Perform multiple imputation approach based on the classical approach (i.e. imputation in a wide format)

   a. Investigate the pattern of missing data. Which variables have missing observations? Are there variables without missing observations?
   b. Specify imputation Equation 4.8 in order to impute missing observations in the proximity scores.
   c. Set the number of imputations to 70 and the inner number of iterations to 10 and perform the imputation task using the FCS (or MICE) algorithm.
   d. Perform the diagnostic tests for the imputed data using all imputations (including iterations).
   e. Prepare the imputed data for the analysis by converting them to long format.
   f. Fit the RI/RS model (Equation 4.1) to the imputed data and correspondingly obtain the pooled results.
   g. Compare the results of part f to those of Table 4.15. Explain possible differences.

5

Analysis of a Pre/Post Measurement Design

5.1 Best Practice for the Analysis of a Pre/Post Measurement Design

5.1.1 The Analysis of Covariance (ANCOVA) Approach

The standard ANCOVA model for a two-group comparison with a baseline measurement as an independent variable is

$$Y_{i2,G} = \beta_0 + \beta_1 G_i + \beta_2 Y_{i1,G} + V_i,$$ (5.1)

where G_i represents the group indicator (coded 0 and 1), $Y_{i1,G}$ and $Y_{i2,G}$ are the pre and post measurements within group G_i, respectively, and V_i is the error term for subject i, $i = 1, \cdots, n$. The errors are independent and normally distributed with mean zero and variance σ_v^2. Also, suppose for the moment that there are no missing observations in any of the variables. The regression parameters of this standard regression model can then be estimated using the OLS method, where the estimate of the group effect can be expressed as:

The regression model in 5.1 implies

$$\overline{Y}_{2,G=0} = \hat{\beta}_0 + \hat{\beta}_2 \overline{Y}_{1,G=0},$$

$$\overline{Y}_{2,G=1} = \hat{\beta}_0 + \hat{\beta}_1 + \hat{\beta}_2 \overline{Y}_{1,G=1},$$

where $\overline{Y}_{k,G=j}$, $j=0, 1$ and $k=1, 2$ are the average of pre and post measurements, respectively. Subtracting both equalities yields

$$\overline{Y}_{2,G=1} - \overline{Y}_{2,G=0} = \hat{\beta}_1 + \hat{\beta}_2 \left(\overline{Y}_{1,G=1} - \overline{Y}_{1,G=0} \right), \text{ or}$$

$$\hat{\beta}_1 = \overline{Y}_{2,G=1} - \overline{Y}_{2,G=0} - \hat{\beta}_2 \left(\overline{Y}_{1,G=1} - \overline{Y}_{1,G=0} \right)$$ (5.2)

DOI: 10.1201/9781003121381-5

Equation 5.2 means that the estimate $\hat{\beta}_1$ is equal to the difference in average at post measurement between two groups but corrected for a fraction of the difference in average at baseline. The estimated value of $\hat{\beta}_2$ is the pooled within-group regression coefficient of the postmeasure on the premeasure (cf. Anderson et al., 1980, p. 250). As a result, the amount of adjustment depends on the correlation between pre and post measurement and on the ratio of post to premeasure variance. In the extreme situation that the correlation is close to zero, hardly any adjustment needs to be made.

A question that may arise is why $\hat{\beta}_1$ rather than the raw difference between the two groups after the intervention ($\overline{Y}_{2,G=1} - \overline{Y}_{2,G=0}$) represents the estimate of the treatment effect. It can be argued that the raw difference may be due to a combination of the intervention effect and group difference at baseline, so that some corrections should be made to estimate the true intervention effect. Considering Equation 5.2 as the preferred choice, there are still additional questions: Is the correction as indicated in Equation 5.2 the correct one? Does Equation 5.2 under all circumstances reflect the true intervention effect or does it only apply under certain conditions? Why not adjust for the whole difference at baseline instead of only a fraction of it?

To adequately answer these questions, a comparison should be made between the corrected difference according to the ANCOVA approach (i.e. Equation 5.2) and the 'true treatment effect'. It is, however, difficult to make such a comparison with the collected data only because the true treatment effect is unknown. Nonetheless, we can still make some general statements, even though the true treatment effect is unknown. To do so, a hypothetical variable is needed that would have been observed in the treatment group had the treatment not been applied. This form of reasoning that we adopt here is known as the 'counterfactual' (Lewis (1973)). It cannot really be observed, but the counterfactual reasoning only serves to oversee the possible bias of the treatment effect in the absence of intervention.

Specifically, suppose $\overline{Y}^*_{2,G=1}$ is the average of a hypothetical post measurement that would have been observed in the treatment group had the treatment not been applied. We emphasise that no data can be collected, as this is a non-existing situation.

Given the counterfactual variable, the treatment effect would be equal to ($\overline{Y}_{2,G=1} - \overline{Y}^*_{2,G=1}$) (See also Figure 5.1), which, in turn, can be expressed by

$$\overline{Y}_{2,G=1} - \overline{Y}^*_{2,G=1} = \overline{Y}_{2,G=1} - \overline{Y}_{2,G=0} - \hat{\beta}^* \left(\overline{Y}_{1,G=1} - \overline{Y}_{1,G=0} \right),$$

with

$$\hat{\beta}^* = \frac{\overline{Y}^*_{2,G=1} - \overline{Y}_{2,G=0}}{\overline{Y}_{1,G=1} - \overline{Y}_{1,G=0}}. \tag{5.3}$$

The reader may check the above equality in Figure 5.1, wherein the dotted line is the hypothetical profile in the treatment group if the treatment had

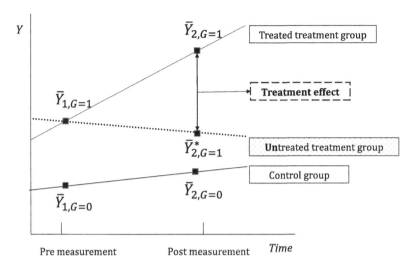

FIGURE 5.1
A graphical representation of the true treatment effect using the counter factual untreated treatment group.

not been applied. The treatment effect is also indicated by means of a double arrow in the figure.

Note that $\hat{\beta}_1$ is an unbiased estimator of the true treatment effect, if the expectation E $(\hat{\beta}_1)$ (notation E (.): average in the population) is equal to the expectation of the counterfactual treatment effect E $(\bar{Y}_{2,G=1} - \bar{Y}_{2,G=1}^*)$.

Thus, $\hat{\beta}_1$ is only an unbiased estimator of the true treatment effect (follows from Equations 5.2 and 5.3), if the following condition is fulfilled.

$$E\left(\hat{\beta}_2\left(\bar{Y}_{1,G=1} - \bar{Y}_{1,G=0}\right)\right) = E\left(\hat{\beta}^*\left(\bar{Y}_{1,G=1} - \bar{Y}_{1,G=0}\right)\right) \tag{5.4}$$

Unfortunately, this condition cannot be tested by the collected data so that it is hard to evaluate whether it is always true in practice.

As discussed in, among others, Anderson et al. (1980, pp. 250) and Weisberg (1979), a randomised clinical trial (RCT) is sufficient for $\hat{\beta}_1$ in Equation 5.2 to be an unbiased estimator. Moreover, it can be shown (see assignment 5.4.7) that the following estimators of β_1 are also unbiased in an RCT:

- The uncorrected difference in post measurement $\hat{\beta}_1 = \bar{Y}_{2,G=1} - \bar{Y}_{2,G=0}$.
- The estimator from the gain – score approach

$$\hat{\beta}_1 = \bar{Y}_{2,G=1} - \bar{Y}_{1,G=1} - \left(\bar{Y}_{2,G=0} - \bar{Y}_{1,G=0}\right).$$

→ *Although the above estimators are unbiased in an RCT, the ANCOVA approach leads to an estimate of the treatment effect with smallest residual variance and largest power (see e.g., Senn, 2006).*

In non-randomised studies, the ANCOVA approach can still deliver unbiased treatment effect estimates, if the treatment assignment is based on the pre-test scores (see van Breukelen, 2013). For example, consider a hypothetical study on the effect of a remedial teaching programme in children with reading difficulties. A reading ability test is administered at baseline and after the remedial teaching. The assignment of children to the remedial teaching programme is not at random, but is based on the baseline scores. Pupils with a high score are assigned to the remedial teaching, while pupils with a low score form the control group. Here, the ANCOVA approach delivers an unbiased estimate of the intervention effect.

> → *Unless a randomised trial is performed or the treatment assignment is based on the pre-test scores, it is unclear whether the ANCOVA approach leads to an unbiased estimate of the treatment effect.*

As mentioned earlier, the estimated correction factor $\hat{\beta}_2$ depends on the correlation between the pre and post measurements. Anderson (1980, p. 250) investigated the correlation between measurements in pre/post measurement designs in educational studies. He found that the correlation between two time points is usually between 0.5 and 0.9, and that the post measurement variance is usually similar to the pre-measurement variance. Using the ANCOVA model in Equation 5.1, the estimated correction factor $\hat{\beta}_2$ will also be between 0.5 and 0.9. Note that $\hat{\beta}_2$ could be larger than one if the variances are unequal, specifically, if the post measurement variance is larger than the pre-measurement variance.

5.1.2 The Gain-Score Approach

In the gain-score analysis, a comparison is made between the change score of both groups. The corresponding estimator of the treatment effect can thus be written as

$$\hat{\beta}_1 = \left(\bar{Y}_{2,G=1} - \bar{Y}_{1,G=1} \right) - \left(\bar{Y}_{2,G=0} - \bar{Y}_{1,G=0} \right) \tag{5.5}$$

In other words, the treatment effect is estimated by the difference between the average change score in the treatment group $(\bar{Y}_{2,G=1} - \bar{Y}_{1,G=1})$ and the average change score in the control group $(\bar{Y}_{2,G=0} - \bar{Y}_{1,G=0})$.

Note that Equation 5.5 can also be rewritten as the difference in average at post measurement between two groups adjusted for the whole difference at baseline (i.e. pre-measurement), that is,

$$\hat{\beta}_1 = \left(\bar{Y}_{2,G=1} - \bar{Y}_{2,G=0} \right) - \left(\bar{Y}_{1,G=1} - \bar{Y}_{1,G=0} \right) \tag{5.6}$$

Equation 5.6 resembles Equation 5.2, but with $\beta_2 = 1$. In analogy to Equation 5.1, the underlying model can thus be specified by

$$Y_{2,G} = \beta_0 + \beta_1 G + 1 \times Y_{1,G} + V. \tag{5.7}$$

Just like in the ANCOVA approach, a relevant question is whether $\hat{\beta}_1$ in Equation 5.6 reflects the correct treatment effect. Because the gain-score analysis implies that $\beta_2 = 1$, it follows from Equations 5.3 and 5.6 that $\hat{\beta}_1$ is an unbiased estimate only if

$$E\left(\bar{Y}_{2,G=1}^* - \bar{Y}_{1,G=1}\right) = E\left(\bar{Y}_{2,G=0} - \bar{Y}_{1,G=0}\right) \tag{5.8}$$

This means that the change over time in the hypothetical treatment group (i.e. in the absence of treatment) is expected to be the same as in the control group. This requirement represents parallel average profiles in the absence of treatment.

> → *It therefore can be concluded that the gain-score analysis implicitly assumes that the control group represents the counterfactual treatment group as far as the change (post- pre) is concerned.*

As a result, the treatment effect would be estimated unbiasedly if this assumption was met. As mentioned before, it cannot be tested by the collected data. Instead, we should rely on content-specific reasoning. In practice, the gain-score approach often means fitting a regression model for the change score (post – pre) on the treatment indicator variable plus prognostic (or confounders) factors if there are any. Two additional properties can be formulated regarding the gain-score approach.

> → *If the treatment assignment is based on pre-test measurement, then gain-score approach delivers biased estimates because it ignores the regression to the mean effect.*
> → *The gain-score approach may produce less biased treatment effect estimates than the ANCOVA approach, if pre-existing groups are involved.*

For instance, suppose a comparison is made between boys and girls in the remedial teaching programme. Here, the gain-score approach may produce less biased estimates than the ANCOVA approach (see, e.g., van Breukelen, 2006 Figure 2).

5.1.3 A Random Intercept Representation of the Gain-Score and ANCOVA Approach

In this section, we demonstrate that the estimate of the treatment effect using a random-intercept model with heterogeneous variances (RI (he)) is identical to that of the gain-score analysis when there are no missing observations. Because the longitudinal model describes the change over time of the difference between the two groups, the specification of the model includes the interaction term *Time × Group*. As a result, the RI (he) model for the pre/post measurement design should be specified as

$$Y_{ij} = \beta_{0i} + \beta_1 G_i + \beta_2 T_{ij} + \beta_3 G_i \times T_{ij} + R_{ij}, \tag{5.9}$$

where G_i and T_{ij} are the group and time variables, respectively, and the error term R_{ij} is uncorrelated and normally distributed with mean zero with heterogeneous variances for subject $i = 1,\ldots, n$ and time point $j = 0, 1$. Both variables G_i and T_{ij} are coded as zero/one such that $G_i = 0$ for untreated group and $G_i = 1$ for treated group, and $T_{ij} = 0$ for before intervention and $T_{ij} = 1$ for after intervention. Note that this RI (he) model is the same as the unstructured marginal model in a pre/post design (see Chapter 4).

To interpret the regression parameters of the fixed part of the model, it makes sense to express their estimates in terms of the average dependent variable values as follows:

$$\hat{\beta}_0 = \bar{Y}_{1,G=0}$$

$$\hat{\beta}_1 = \bar{Y}_{1,G=1} - \bar{Y}_{1,G=0}$$

$$\hat{\beta}_2 = \bar{Y}_{2,G=0} - \bar{Y}_{1,G=0}$$

$$\hat{\beta}_3 = \left(\bar{Y}_{2,G=1} - \bar{Y}_{2,G=0}\right) - \left(\bar{Y}_{1,G=1} - \bar{Y}_{1,G=0}\right) \tag{5.10}$$

Suppose there are n_1 subjects in the control group (i.e., when G=0) and n_2 subjects in the experimental group (i.e., when G=1).

Using the well-known results of the regression analysis and by filling in the codes of independent variables in Equation 5.9, it can be seen that

$$\bar{Y}_{j,G=0} = \hat{\beta}_0 + \hat{\beta}_2 T_{j,G=0}, \text{ where } \bar{Y}_{j,G=0} = \frac{1}{n_1}\sum_{i=1}^{n_1} Y_{ij}, T_{j,G=0} = \frac{1}{n_1}\sum_{i=1}^{n_1} T_{ij}, \text{ and}$$

$$\bar{Y}_{j,G=1} = \hat{\beta}_0 + \hat{\beta}_1 + \hat{\beta}_2 T_{j,G=1} + \hat{\beta}_3 T_{j,G=1}, \text{ where } \bar{Y}_{j,G=1} = \frac{1}{n_2}\sum_{i=1}^{n_2} Y_{ij}, T_{j,G=1} = \frac{1}{n_2}\sum_{i=1}^{n_2} T_{ij}.$$

Substituting the codes for baseline (T=0) and post measurement (T=1) into the above equalities yields

$$\bar{Y}_{2,G=0} = \hat{\beta}_0 + \hat{\beta}_2$$

$$\bar{Y}_{2,G=1} = \hat{\beta}_0 + \hat{\beta}_1 + \hat{\beta}_2 + \hat{\beta}_3$$

$$\bar{Y}_{1,G=0} = \hat{\beta}_0$$

$$\bar{Y}_{1,G=1} = \hat{\beta}_0 + \hat{\beta}_1$$

Equation 5.10 follows by expressing the estimated β's in terms of the average of Y's.

In the set of equations mentioned in 5.10, each estimate of β can be interpreted in terms of the observed average of the dependent variable per group and per time point. Moreover, it can be seen that $\hat{\beta}_3$ in Equation 5.10 is identical to Equation 5.6, which reflects the difference in gain scores between the two groups.

→ *Hence, it can be concluded that in a pre/post measurement design, the gain-score approach can be represented by an RI (he) model that includes the interaction between group and time.*

This can also be demonstrated by means of an example. Consider the first two repeated measurements of the Growth study. Tables 5.1a and 5.1b show the results of a random-intercept and the gain-score analysis, respectively. Note that the RI (he) model has *Distance* as the dependent variable (cf. Equation 5.9) with group $G = Sex$ and time $T = Age$, whereas the gain-score approach has the change in *Distance* as the dependent variable (cf. Equation 5.7). From these tables, it appears that not only the point estimate of the interaction term in the random-intercept model is identical to the point estimate

TABLE 5.1A

Growth Study: REML Estimates of an RI (he) Model for the First Two Measurements

Parameter	Estimate	Std. Error	df	t-value	p-value	95% Confidence Interval	
	Lower Bound	Upper Bound	Lower Bound	Upper Bound	Lower Bound	Upper Bound	Lower Bound
Intercept	22.87	0.55	37.87	41.76	0.00	21.77	23.98
Sex	−1.69	0.86	37.87	−1.97	0.05	−3.43	0.04
Age	0.94	0.51	25	1.84	0.08	−0.11	1.99
Sex × Age	0.11	0.80	25	0.14	0.89	−1.54	1.75

Dependent Variable: *Distance.*

TABLE 5.1B

Growth Study: The Gain-Score Approach for the First Two Measurements

Parameter	Estimate	Std. Error	df	t-value	p-value	95% Confidence Interval	
	Lower Bound	Upper Bound	Lower Bound	Upper Bound	Lower Bound	Upper Bound	Lower Bound
Intercept	0.94	0.51	25	1.84	0.08	−0.11	1.99
Sex	0.11	0.80	25	0.14	0.89	−1.54	1.75

Dependent Variable: *Change.*

of the *Sex* coefficient in the gain-score approach, but their corresponding standard errors and the 95% confidence intervals are also the same.

It can also be shown that the ANCOVA approach for testing the treatment effect (as described in Section 5.1.1) is equivalent to testing the interaction effect β_3 in Equation 5.11 (see, van Breukelen 2013). This model is identical to the RI (he) Equation 5.9 with a restriction that sets the group effect β_1 to zero because the ANCOVA approach assumes that there is no group difference at pre-measurement (Holland and Rubin, 1983; Rubin, 1974, 1977; van Breukelen, 2013).

$$Y_{ij} = \beta_{0i} + \beta_2 T_{ij} + \beta_3 T_{ij} \times G_i + R_{ij}, \tag{5.11}$$

We will denote this model as the 'restricted RI (he)' model.
In conclusion, we summarise our findings as follows:

- *In RCT, the ANCOVA and gain-score approach are both correct, in the sense that the treatment effect is unbiased. However, the ANCOVA approach is the most appropriate method because it gives a smaller standard error of the estimated treatment effect and hence a larger power and higher precision.*
- *In quasi-experimental and life-event studies, the ANCOVA approach will yield unbiased results if the assignment to the groups is based on the baseline scores (Senn, 2006, van Breukelen, 2006, 2013). The gain-score approach, however, delivers biased estimates because it ignores the regression to the mean effect.*
- *The gain-score approach can yield less biased estimates than the ANCOVA approach in studies wherein pre-existing groups are compared (e.g. men vs. women).*

5.2 Case Studies

5.2.1 An Example of a Randomised Clinical Trial: Beating the Blues

Beating the Blues (BtB, Proudfoot et al., 2003) was a clinical trial designed to assess the effectiveness of an interactive programme using multi-media techniques for a cognitive behavioural therapy of depressed patients. In this study, patients with depression recruited in primary care were randomised to either the BtB program, or to the Treatment as Usual program (TAU). The variable 'Treat' represents two treatments (*Treat* = 1 if BtB, and *Treat* = 0 if TAU). The dependent variable (*Depress*) used in the trial was the Beck Depression Inventory II, with higher values indicating more depression. Measurements of this variable were made on five occasions, which are as follows:

- *prior to treatment (Bdipre), and*
- *follow up at 2, 3, 5 and 8 months after treatment.*

The scientific question of interest was whether the BtB programme performs better than the TAU in treating depression.

In this section, we consider the *Depress* score at baseline and two months after the treatment to represent a pre/post measurement study. There were three subjects in the TAU group who had missing observations at post measurement. These subjects were thus excluded for further analysis. The analysis in the presence of missing observations will be investigated in Section 5.3.

We use the data that was reported in the study by Landau and Everitt (2004) for the analysis. To demonstrate the results from different approaches (the ANCOVA, gain-score and RI (he) model), we start with descriptive statistics presented in Table 5.2.

The raw difference in average depression score at the pre-measurement is $\bar{Y}_{pre,BTB} - \bar{Y}_{pre,TAU} = -1.33$. Likewise, the raw difference in averages depression score at the post measurement is $\bar{Y}_{post,BTB} - \bar{Y}_{post,TAU} = -4.76$.

Figure 5.2 shows these raw average depression scores at pre and post measurements within each treatment arm. The raw difference average at post measurement will be corrected by a fraction of the difference in averages at baseline (i.e. pre-measurement).

According to the ANCOVA approach (Equation 5.2),

$$\hat{\beta}_1 = \bar{Y}_{post,BTB} - \bar{Y}_{post,TAU} - \hat{\beta}_2 \left(\bar{Y}_{pre,BTB} - \bar{Y}_{pre,TAU} \right) = -3.96,$$

TABLE 5.2

Beating the Blues Study: Descriptive Statistics for the Pre and Post Measurement of Depression Score by Treatment Arm

Treatment		N	Minimum	Maximum	Mean	Std. Deviation
TAU	Depression at baseline ($\bar{Y}_{pre,TAU}$)	45	7	47	23.87	9.65
	Depression two months after treatment ($\bar{Y}_{post,TAU}$)	45	0	48	19.47	11.08
BtB	Depression at baseline ($\bar{Y}_{pre,BTB}$)	52	2	49	22.54	11.74
	Depression two months after treatment ($\bar{Y}_{post,BTB}$)	52	0	40	14.71	10.12

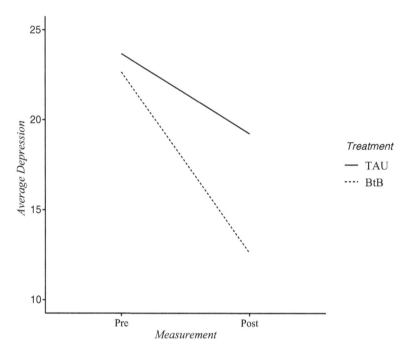

FIGURE 5.2
Raw averages of *Depression* at pre and post measurements for each treatment group in the Beating the Blues study.

which is the same as the estimated adjusted treatment effect in Table 5.3a (the difference in the second decimal is due to rounding error). It should be noted that $\hat{\beta}_2 = 0.60$ is extracted from Table 5.3a and not from Table 5.2 for the above calculation. According to the change score approach (Equation 5.6),

$$\hat{\beta}_1 = \left(\bar{Y}_{post,BTB} - \bar{Y}_{post,TAU}\right) - \left(\bar{Y}_{pre,BTB} - \bar{Y}_{pre,TAU}\right) = -3.43,$$

which is the same as the estimated treatment effect using the standard regression analysis presented in Table 5.3b.

Following the RI (he) model (Equation 5.9), the adjusted treatment effect is estimated by the regression coefficient of the interaction term in Table 5.3c, which is equal to the estimated treatment effect in the gain-score analysis.

Using the ANCOVA approach, on the one hand, the estimated treatment effect is significant at the 5% level ($\hat{\beta}_1 = -3.95$ (std.error = 1.71), p-value = 0.023). The gain-score approach, on the other hand, shows a nonsignificant treatment effect estimate at the 5% level ($\hat{\beta}_1 = -3.43$ (std.error = 1.91), $p=0.076$). The discrepancy is because the gain-score approach leads to a smaller treatment effect with a larger standard error; that is, the post treatment raw difference is adjusted for the whole raw difference at baseline, whereas in the ANCOVA approach, the adjustment is made only for a fraction of the raw difference at baseline.

The results from the ANCOVA approach can also be obtained by fitting the restricted RI (he) model (see Equation 5.11) to the data. The corresponding results are depicted in Table 5.3.d, wherein the estimate of the interaction term resembles the estimated treatment effect in the ANCOVA model (also the 95% confidence interval and *p*-value). It should nevertheless be noted that the estimates of the treatment effect from the ANCOVA and restricted

TABLE 5.3A

Beating the Blues Study: Results of the ANCOVA Approach

Parameter	Estimate	Std. Error	df	*t*-value	*p*-value	95% Confidence Interval	
						Lower Bound	Upper Bound
Intercept	5.08	2.27	94	2.24	0.027	0.58	9.58
Treat	−3.95	1.71	94	−2.32	0.023	−7.34	−0.57
Bdipre	0.60	0.08	94	70.60	0.000	0.45	0.765

Dependent Variable: *Depression*.

TABLE 5.3B

Beating the Blues Study: Results of the Gain-Score Approach

Parameter	Estimate	Std. Error	Df	*t*-value	*p*-value	95% Confidence Interval	
						Lower Bound	Upper Bound
Intercept	−4.40	1.40	95	−3.15	0.002	−7.17	−1.63
Treat	−3.43	1.91	95	−1.80	0.076	−7.21	0.36

Dependent Variable: *Change*.

TABLE 5.3C

Beating the Blues Study: Results of the RI (he) Model

Parameter	Estimate	Std. Error	df	*t*-value	*p*-value	95% Confidence Interval	
						Lower Bound	Upper Bound
Intercept	23.87	1.61	95	14.79	0.000	20.66	27.07
Treat	−1.33	2.20	95	−0.60	0.548	−5.70	3.05
Time	−4.40	1.40	95	−3.15	0.002	−7.17	−1.63
Treat × Time	−3.43	1.91	95	−1.80	0.076	−7.21	0.36

Dependent Variable: *Depression*. The REML estimation method is used.

TABLE 5.3.D

Beating the Blues Study: Results of the Restricted RI (he) Model

Parameter	Estimate	Std. Error	df	t-value	p-value	95% Confidence Interval Lower Bound	Upper Bound
Intercept	23.15	1.10	96	21.14	0.000	20.98	25.33
Time	−4.13	1.31	106.17	−3.13	0.002	−6.72	−1.51
Treat × Time	−3.95	1.69	95	−2.33	0.022	−7.32	−0.59

Dependent Variable: depression score. The REML estimation method is used.

RI (he) model are not identical so that small differences can be expected. The difference becomes more pronounced as the sample size decreases (van Breukelen, 2013).

Finally, we emphasise that the ANCOVA approach has more power for randomised clinical trials. The ANCOVA approach is thus the preferred choice given that the randomisation has been done properly in this study.

5.2.2 An Example of a Nonexperimental Study: Well-Being, a Life-Event Study

A study in which the impact of a certain event such as illness or loss of a partner is investigated is called a life-event study with a 'natural' intervention. Such a live-event study was reported by Nieboer et al. (1998), who studied the effect of loss and illness of a partner on the state of well-being for 269 respondents. There were 157 caregivers whose partner had an illness-incidence and 112 widow(er)s whose partner died. The variable *Group* distinguishes between the groups (0 for caregivers and 1 for widow(re)s). Three repeated measurements were taken on *Well-being* (a quantitative variable; the higher the score, the better the well-being): a baseline measurement, a post measurement after three months and a follow-up measurement after 12 months. The variable *Gender* identifies the respondent's gender (0 for males and 1 for females). The quantitative variable *Age* is a possible confounder.

The main objective of the study was to compare caregivers and widow(er)s and males and females with respect to their *well-being* after the life event. For the sake of simplicity, we focus on the difference in average well-being between males and females (gender effect) due to the life event and consider the baseline measurement and the post measurement after three months only. Moreover, we restrict the analysis to participants for whom both the pre and post measurement were observed. The case of missing observations is discussed in the next section. Table 5.4 shows descriptive statistics for *Well-being* at baseline and three months after the live event.

TABLE 5.4

Well-Being Study: Descriptive Statistics for the Dependent Variable by Gender

Respondents' Gender		N	Minimum	Maximum	Mean	Std. Deviation
Male	*Well- being* at baseline ($\bar{Y}_{pre,male}$)	66	11.00	36.00	27.02	3.57
	Well-being after three months ($\bar{Y}_{post,male}$)	66	8.00	31.00	24.08	5.82
Female	*Well- being* at baseline ($\bar{Y}_{pre,female}$)	135	10.00	35.00	25.81	4.59
	Well-being after three months ($\bar{Y}_{post,female}$)	135	3.00	31.00	20.57	6.26

The raw difference in average *Well-being* at the baseline is $\bar{Y}_{pre,female} - \bar{Y}_{pre,male} = -1.21$. Likewise, the raw difference in average *Well-being* at the post measurement is $\bar{Y}_{post,female} - \bar{Y}_{post,male} = -3.51$. Figure 5.3 depicts these raw averages at baseline and post measurement for males and females separately.

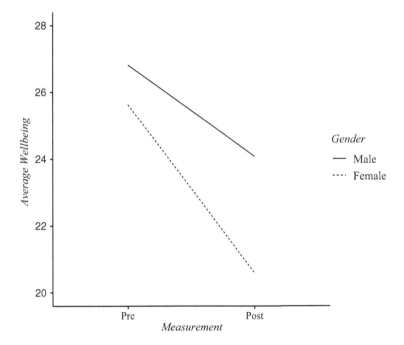

FIGURE 5.3

Raw averages of *Well-being* at pre and post measurements for each gender group in the Well-being study with only complete cases.

According to the ANCOVA approach (Equation 5.2),

$$\hat{\beta}_1 = \left(\bar{Y}_{post,female} - \bar{Y}_{post,male}\right) - \hat{\beta}_2\left(\bar{Y}_{pre,female} - \bar{Y}_{pre,male}\right) = -2.84,$$

which is equal to the estimated effect of *Gender* (due to the life event) in Table 5.5a. Note that the restricted ANCOVA in Table 5.5d also results in the same gender effect estimate. According to the change score approach (equation 5.6),

$$\hat{\beta}_1 = \left(\bar{Y}_{post,female} - \bar{Y}_{post,male}\right) - \left(\bar{Y}_{pre,female} - \bar{Y}_{pre,male}\right) = -2.30,$$

which is equal to the estimated regression slope for *Gender* (see, Table 5.5b).

According to the RI (he) model (Equation 5.9), the adjusted effect of *Gender* is estimated by the regression coefficient of the interaction term in Table 5.5c, which is identical to the estimated gender effect in the gain-score analysis.

In theory, the gain-score approach gives less unbiased results than the ANCOVA approach because of the comparison between pre-existing groups such as *Gender*. However, all analyses confirm that the male participants differ significantly from the female participants in *Well-being* three months after the life event, adjusted for baseline *Well-being* by either method. Although

TABLE 5.5A

Well-Being Study: Results of the ANCOVA Approach

Parameter	Estimate	Std. Error	df	*t*-value	*p*-value	95% Confidence Interval Lower Bound	95% Confidence Interval Upper Bound
Intercept	9.15	2.62	198	3.49	0.000	3.97	14.32
Well-being at baseline	0.55	0.09	198	5.90	0.000	0.37	0.74
Gender	−2.84	0.86	198	−3.31	0.001	−4.53	−1.15

Dependent Variable: *Well-being*.

TABLE 5.5B

Well-Being Study: Results of the Gain-Score Approach

Parameter	Estimate	Std. Error	df	*t*-value	*p*-value	95% Confidence Interval Lower Bound	95% Confidence Interval Upper Bound
Intercept	−2.94	0.73	199	−4.01	<0.001	−4.39	−1.49
Gender	−2.30	0.89	199	−2.57	0.011	−4.06	−0.53

Dependent Variable: *Change*.

TABLE 5.5C

Well-Being Study: Results of the RI (he) Model

Parameter	Estimate	Std. Error	df	t-value	p-value	95% Confidence Interval Lower Bound	Upper Bound
Intercept	27.02	0.53	199	51.27	<0.001	25.98	28.05
Time	−2.94	0.73	199	−4.01	<0.001	−4.39	−1.49
Gender	−1.21	0.64	199	−1.88	0.062	−2.48	0.06
Time × Gender	−2.30	0.89	199	−2.57	0.011	−4.06	−0.53

Dependent Variable: *Well-being*. The REML estimation method is used.

TABLE 5.5D

Well-Being Study: Results of the Restricted RI (he) Model

Parameter	Estimate	Std. Error	df	t-value	p-value	95% Confidence Interval Lower Bound	Upper Bound
Intercept	26.20	0.303	200	86.25	0.000	25.60	26.80
Time	−2.58	0.71	208.91	−3.64	<0.001	−3.97	−1.18
Time × Gender	−2.84	0.85	199	−3.35	<0.001	−4.51	−1.17

Dependent Variable: *Well-being*. The REML estimation method is used.

the magnitude of the estimated effects is different between the ANCOVA and gain-score analysis, the conclusion remains the same. In sum, we conclude that both approaches lead to the same overall conclusion in this nonexperimental example. Nonetheless, this may not always be the case.

5.3 Handling Missing Observations in Pre/Post Measurement Designs

In the previous sections, we assumed that the data do not have missing observations and reviewed two approaches to analyse data from the pre/post measurement designs. This section expands the analysis to the case of a pre/post measurement design in the presence of missing observations. We discuss the consequence of incompleteness of the data when the ANCOVA or gain-score approach is used and compare different methods of dealing with the missing data. We further demonstrate through case studies how a sensitivity analysis can be performed in pre/post measurement designs.

5.3.1 The ANCOVA Approach

Focusing on a pre/post measurement design, suppose the regression model of interest is

$$Y_{i2,G} = \beta_0 + \beta_1 G_i + \beta_2 Y_{i1,G} + V_i \qquad (5.12)$$

where i indexes participant, Y_{i1} and Y_{i2} are pre and post measurement scores respectively, G_i is a grouping variable representing two distinct groups (coded as 0 and 1) such as married versus unmarried subjects or treated versus untreated (e.g. control) patients, and $V_i \sim N(0, \sigma_v^2)$. Note that Equation 5.12 is identical to Equation 5.1, and the main interest is to compare two groups with respect to the post measurement while accounting for the pre-measurement (i.e. baseline).

Suppose further the intended pre or post measurement is not obtained for all participants so that the standard ANCOVA approach (Equation 5.12) implies using complete cases only; that is, participants for which both the pre and post measurements are fully observed are included in the analysis. Unlike the case of no missing data, the estimate of the treatment effect using the ANCOVA analysis is not necessarily equivalent to that of the restricted random-effects models introduced in Section 5.1.2 because the former relies on complete cases for the analysis, while the latter makes use of available cases. Therefore, the difference in parameter estimates between the ANCOVA and restricted random-effects model is evident, as the methods do not use the same source of information.

→ *It can thus be concluded that, the ANCOVA (equation 5.12) and restricted random-effects model (equation 5.11) do not necessarily produce equivalent estimates of treatment effects in pre/post measurement designs with missing observations.*

What implications does this difference have for the ANCOVA and restricted random-effects model? They both can be valid and produce unbiased estimates of the treatment effect if the analysis Equation 5.12 is correct and the missing data mechanism is unrelated to the post measurement (see Tables 2.12 and 3.13). Nonetheless, some precision is generally lost under ANCOVA compared with the restricted random-effects model, as the latter uses more information. Specifically, the restricted random-effects model can be superior to ANCOVA when the pre-measurement or both the pre and post measurements are partially observed. The reason is that some post measurements will be excluded when the ANCOVA is applied, while the restricted random-effects model still uses those measurements in the analysis. The superiority of the restricted random-effects model seems to vanish when missing observations are limited to the post measurement only, as both methods are equally efficient (see example below).

We elaborate on this rather theoretical and abstract explanation and compare the different methods using the Beating the Blues trial. In this study, two treatment groups (BtB vs. TAU) are compared with respect to the patient's depression score. The study investigates the change in average of depression from baseline to the study end (i.e. 8 months after the treatment) between two treatment groups. The analysis model is thus Equation 5.12, which is the regression of depression score at month eight on treatment, adjusted for baseline depression score. Table 5.6 describes the summary statistics for depression at pre and post measurement, stratified by treatment group.

In this study, there were no missing observations at baseline (i.e. pre-measurement), while about 50% of patients (48 out of 100) did not have the post measurement score at the end of the study. For the moment, we assume that the missing data mechanism is unrelated to the post measurement, or specifically, the missing data mechanism for post measurement is MAR conditional on treatment assignment and/or pre-measurement. This means that the probability of missingness in the post measurement is assumed equal for all patients who are in the same treatment group and have the same pre-measurement score.

The estimate of the treatment effect β_1 from fitting the ANCOVA model (Equation 5.12) to complete records is shown in the first row of Table 5.7.

TABLE 5.6

Beating the Blues Study: Summary Statistics for Depression Score, Stratified by Treatment Group

Month 8 Status	Treatment	n	Pre-measurement Mean (SD)	Post Measurement Mean (SD)
Observed		52		
	TAU	25	24.1 (8.07)	13.6 (11.50)
	BtB	27	22 (10.90)	8.8 (6.10)
Missing		48		
	TAU	23	24.3 (11.60)	?
	BtB	25	23.1 (12.80)	?

TABLE 5.7

Beating the Blues Study: Estimates of the Treatment Effect at the End of the Study, Adjusted for Baseline

Method	Estimate	SE	95% CI
ANCOVA	−4.01	2.38	(−8.79, 0.77)
MI, $m = 10{,}000$	−4.00	2.42	(−8.78, 0.79)
Restricted RI (he)	−4.01	2.34	(−8.71, 0.69)

All methods are evaluated under the MAR assumption. MI stands for multiple imputation with m = 10,000 imputations. Restricted RI (he) is Equation 5.11.

We also apply multiple imputation (MI), where missing post measurements are imputed using the pre-measurement and treatment allocation. This procedure essentially assumes the MAR mechanism by default. In line with theory, the results of MI hardly deviate from complete case analysis (second row of Table 5.7) if the number of imputations is very large. This means that there is no advantage to, or gain from, using MI in this simple setting. Finally, the restricted RI (he) model (Equation 5.11) is fitted to the data. As expected, all methods agree very closely showing no statistically significant difference between the treatment groups at the end of the study (at 5% level). We may hence conclude that all methods are comparable when missing observations are confined to the dependent variable. Nevertheless, the methods do not necessarily deliver equivalent results when missing observations extend to pre and post measurements.

Generally, it is very hard to justify the MAR assumption using the data at hand. The main reason relates to the fact that the observed data cannot confirm dependency of missingness to the observed data nor disprove it (see, Chapter 1, Section 1.3.1 for an illustration).

> → *Consequently, any inference from incomplete data should start with an analysis under MAR and subsequently investigates the robustness of inference to departure from the MAR assumption.*

As formally introduced in Chapter 3 (Section 3.5.4), various ways exist to conduct sensitivity analysis. Interested readers may consult Little and Rubin (2020, chap. 15) and Molenberghs et al. (2015, section V) and references therein for recent developments in this area. In this study, we opt for the sensitivity analysis with MI because it is easy to understand the nature of sensitivity parameters and to implement the procedure in standard software.

Before formally framing the sensitivity analysis, it is worth to discuss the plausibility of the MAR assumption in practice. We started with a MAR assumption in the Beating the Blues trial, where the missingness of post measurement could depend on the treatment allocation and/or on the pre-measurement. This, however, is not the only possible MAR mechanism. If additional variables like auxiliary variables or secondary dependent variables are available, these variables could explain, to some extent, the missingness of the post measurement. For example, it is known that a history of depression might have a relation with dropout, that is patients with a (long) history of depression are more likely to dropout than others. Hence, incorporation of such variables in the analysis can make the MAR assumption more plausible. Although such additional information can directly be incorporated into the direct likelihood methods, MI offers a more convenient solution because these additional variables can be used in the imputation phase while excluded from the final analysis making the main analysis consistent with the study protocol. Note that the analysis that incorporates additional information still makes the MAR assumption but is not formally a part of the sensitivity analysis. Nevertheless, it still helps to understand the value of additional information to the inference.

Sensitivity analysis to explore the robustness of inferences to the MAR assumption typically involves the selection and pattern mixture models (see Chapter 3). Here, we focus on the pattern mixture models, as multiple imputation can conveniently be adapted to accommodate the pattern mixture framework. To understand the framing of sensitivity analysis, let us consider the Beating the Blues study as a case study. Furthermore, suppose Equation 5.12 is the analysis model with post measurement of depression at month 8 as the dependent variable and treatment allocation and pre-measurement of depression as the independent variables. Because the pre-measurement scores are fully observed, there are only two patterns with respect to missingness (see Table 5.6):

- *Pattern 1: pre and post measurements are both observed*
- *Pattern 2: pre-measurement is only observed*

We also analyse the data based on the intention-to-treat principle, that is comparing patients in the groups to which they were originally randomly assigned, regardless of the treatment they received.

Using the pattern mixture framework, the ANCOVA model implies that Equation 5.12 is specified for the first pattern. Similarly, the second pattern can be defined, for example, by

$$Y_{i2,G} = \alpha_0 + \alpha_1 G_i + \alpha_2 Y_{i1,G} + U_i \tag{5.13}$$

where α's are the regression weights and $U_i \sim N(0, \sigma_u^2)$. Ideally, missing post measurements should be imputed using Equation 5.13. However, the model parameters (i.e. α_0, α_1, α_2 and σ_u^2) are unidentified because $Y_{i2,G}$ is entirely unobserved. As a result, identifying restrictions should be placed on these parameters. For instance, by assuming that the parameters of Equation 5.13 are identical to their counterparts in model Equation 5.12, we can impute missing post measurements using Equation 5.13, while the parameters are obtained from Equation 5.12.

It should be emphasised that the standard implementation of MI (under MAR) implicitly makes this equality assumption. It therefore can be concluded that the MAR assumption within the MI framework means both Equation 5.12 and Equation 5.13 are identical. The equality assumption, however, cannot be tested from the observed data simply because $Y_{i2,G}$ is unavailable in Equation 5.13. We thus stress the necessity of conducting sensitivity analysis when data involve missing observations.

To explore deviations from the MAR assumption, a natural way is to rewrite the model with Equation 5.13 as an extension of the model with Equation 5.12, that is,

$$Y_{i2,G} = \left(\beta_0 + \delta_0\right) + \left(\beta_1 + \delta_1\right)G_i + \left(\beta_2 + \delta_2\right)Y_{i1,G} + \lambda V_i, \qquad (5.14)$$

where $\alpha_0 = (\beta_0 + \delta_0)$, $\alpha_1 = (\beta_1 + \delta_1)$, $\alpha_2 = (\beta_2 + \delta_2)$, and $U_i = \lambda V_i$. Note that Equation 5.14 reduces to Equation 5.12 by fixing $\delta_0 = \delta_1 = \delta_2 = 0$, and $\lambda = 1$. These unknown parameters are called *sensitivity parameters*, and they cannot be identified or estimated because there is no information about them in the data. The main advantage of specifying sensitivity parameters as mentioned above is transparency since such fixed transformations are easy to create and communicate. This point was advanced by Rubin (1987, Chap. 6), among others, stressing that a successful sensitivity analysis requires a clear expression of the implied assumptions so that their relevance and plausibility can be identified. For the ease of presentation, we fix $\delta_2 = 0$ and $\lambda = 1$ and use specific rules and information to specify plausible values for δ_0 and δ_1 in the Beating the Blues example. It is not uncommon to assume that missing observations of post measurement have a different, usually poorer, response than those missing observations predicted under MAR. In the Beating the Blues study, as an example, the depression score might hence decline less quickly after withdrawal within one or both groups. This implies that if missing observations had been observed, these could have been higher than the observed ones. Figure 5.4 illustrates such a hypothetical situation, wherein BtB patients with missing scores could have higher scores at post measurement compared with patients with observed scores.

The black and grey solid lines are the observed profiles for TAU and BtB groups, respectively. The dotted line shows that the profile for BtB group could have increased had the missing observations been observed. Note that the dotted and grey solid lines coincide under the MAR assumption.

Focusing on the above scenario, the main question considers choosing plausible values for the sensitivity parameters δ_0 and δ_1. The models with Equation 5.12 and Equation 5.14 imply that, at post measurement, δ_0 and $\delta_0 + \delta_1$ are the mean difference in depression score between pattern 1 and pattern 2 in the TAU and BtB groups, respectively. From Table 5.6, it turns out that these mean differences at baseline were close to zero (24.1 vs. 24.3 in TAU group) and one (22 vs. 23.1 in BtB group), respectively. Consequently, we may expect a similar pattern at post measurement and choose zero and one as plausible values for δ_0 and δ_1

Table 5.8 shows the results of sensitivity analysis for various combination of δ_0 and δ_1. Here, we

1) Impute missing post measurements under MAR using Equation 5.12, and then

2) The imputed values are amended according to the assigned values for δ_0 and δ_1, resulting in imputations under the posited MNAR mechanism.

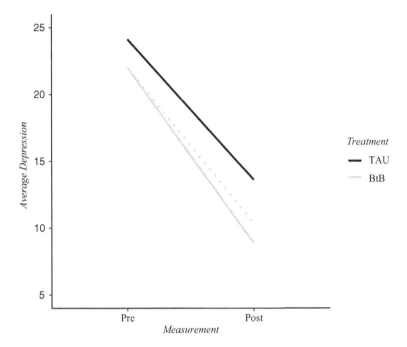

FIGURE 5.4
A schematic illustration of an increasing observed profile (dotted line in the BtB arm of the Beating the Blues study under the posited MNAR mechanism.

The abovementioned steps are then repeated to create each imputed dataset, wherein the parameter(s) of interest are estimated from each imputed dataset in turn. The results are then pooled to obtain a single inference.

The first row of Table 5.8 shows the treatment effect estimate under MAR from multiple imputation with m = 1000 (which only takes a matter of minutes in this example). In all MNAR scenarios, the treatment effect estimate decreased, but it remains insignificant at 5% level. The precision is also unchanged because the posited MNAR model only shifts the distribution of missing observations but not the variance. Finally, we conclude that our analysis of Beating the Blues study is robust to the MAR assumption after a sensitivity analysis under the assumed MNAR mechanism.

This section closes with paying special attention to the final points:

- In the Beating the Blues example, we assumed an increasing profile for missing observations of the post measurement. A decreasing profile could also be considered (if plausible) by assigning relevant values to the sensitivity parameters.
- We have performed MI with sensitivity analysis without incorporating any auxiliary variables. If such information is available, the

TABLE 5.8

Beating the Blues Study: Estimates of the Treatment Effect from Multiple Imputation under MAR and MNAR with Different Values for Sensitivity Parameters δ_0 and δ_1

Mechanism	Estimate	SE	T-value	95% CI
MAR	−4.001	2.395	−1.670	(−8.815, 0.814)
MNAR				
$\delta_0 = 0, \delta_1 = 0.5$	−3.759	2.396	−1.569	(−8.574, 1.056)
$\delta_0 = 0, \delta_1 = 1$	−3.518	2.397	−1.468	(−8.334, 1.299)
$\delta_0 = 0, \delta_1 = 1.5$	−3.276	2.398	−1.366	(−8.096, 1.543)
$\delta_0 = 1, \delta_1 = 1$	−3.514	2.401	−1.464	(−8.339, 1.311)
$\delta_0 = 1, \delta_1 = 0.5$	−3.755	2.399	−1.565	(−8.577, 1.066)
$\delta_0 = 1, \delta_1 = 0$	−3.997	2.398	−1.667	(−8.815, 0.822)

MI analyses used m = 1000 imputations. The analysis model (Equation 5.12) is used to estimate parameters after MI.

whole procedure can be adapted accordingly, that is the MI under MAR uses these variables in the imputation phase, and subsequently, the imputed values are adjusted using the sensitivity parameters. For instance, the history of depression needs be added to the imputation model if it relates to the dropout process. The imputed values are then amended using the assigned values of the sensitivity parameters.

- We have made an educated guess about plausible values of the sensitivity parameters in our example. Another possibility is to elicit plausible values for the sensitivity parameters from experts (see, White et al., 2007 for an illustration).

- Using the Beating the Blues trial as a case study, we have only covered the missing data problem in the dependent variable. When, however, the independent variables also have missing observations, the sensitivity analysis should be extended accordingly.

- In experimental studies like randomised controlled trials (RCTs), missing observations of pre-measurement do not cause bias in the treatment effect estimate because their missingness is unrelated to the post measurement scores (see, Table 2.12). This is because it is highly implausible to assume that the missingness of pre-measurement depends on some other measurements that will be obtained in the future (logical fallacy). In this very specific case, the missing values of pre measurement can be imputed deterministically using the 'mean imputation' approach (see, for a detailed discussion, White and Thompson, 2005; Kayembe et al., 2020; Kayembe et al. 2022a), while the missing values of the post measurements should be imputed as shown in this section.

5.3.2 The Gain-Score Approach

The issue of missing data using the gain-score approach is basically not different from that of the ANCOVA approach. Here, on the one hand, gain scores are obtained from completers, that is subjects for which both the pre and post measurements are observed, and then the standard linear regression on the group variable or the t-test is used to compare (intervention) the groups. Applying the random-effects models, on the other hand, implies using the available cases (see also assignment 5.5.6). Like the ANCOVA approach, it can hence be concluded that

→ *The estimate of the treatment effect in the gain-score approach (equation 5.7) is not necessarily identical to that of the random-effects model (equation 5.9) in the presence of missing data in pre/post measurement designs.*

Imputation of missing observations and the subsequent sensitivity analysis for the gain-score approach can also be performed in a similar manner as in the ANCOVA approach. While we used the wide data format for imputation in the ANCOVA approach, we take another path in the gain-score approach and impute missing observations from the long data format. Therefore, imputations are generated using the developed multilevel imputation methodology in Chapter 4.

To begin with, let us revisit the Well-being study introduced in Section 5.3. Consistent with the analysis that was preformed earlier, we focus on the comparison between males and females on change score in well-being from baseline to three months after the life event. Table 5.9 summarises the patterns of missing data in this study. Seventy-five percent of individuals were completers, while the remaining did not provide the well-being measurement in one or both occasions.

Suppose the analysis model of interest is a random-intercept model (see also equation 3.5) taking the form:

$$Y_{ij} = \beta_0 + \beta_1 G_i + \beta_2 T_{ij} + \beta_3 G_i \times T_{ij} + G_{0i} + R_{ij}, \tag{5.15}$$

TABLE 5.9

Well-Being Study: Patterns of Missing Data

Pattern Group	Gender	Age	Group	Well-being Measurement Pre	Post	Observed #	%
1	O	O	O	O	O	201	75%
2	O	O	O	O	M	63	23%
3	O	O	O	M	O	3	1%
4	O	O	O	M	M	2	1%
Missing	0(0%)	0(0%)	0(0%)	5(2%)	65(24%)		

where i and j index participants and time points ($j = 0, 1$), so that Y_{i0} and Y_{i1} represent the well-being scores at baseline and three months after the life event, respectively. The group variable G_i is coded as zero (for males) and one (for females), and $T_{ij} = 0$ at baseline (i.e, pre-measurement) and $T_{ij} = 1$ at post measurement. Finally, the random components G_{0i} and R_{ij} are independent and $G_{0i} \sim N(0, \tau_0^2)$, $R_{ij} \sim N(0, \sigma^2)$. In this model, the interaction term β_3 is of main scientific interest because it quantifies the difference in well-being at post measurement between males and females after correcting for its baseline difference.

Starting with the MAR assumption, the direct likelihood approach implies fitting the random-intercept model (Equation 5.15) to the data. Here, two subjects (in pattern 4) are excluded from the analysis because these units do not have any score for the pre and post measurements. Next, we apply MI to the data, where the fully observed independent variables age and group (i.e. widowers vs. caregivers) are added to the imputation model to make the MAR assumption more plausible.

Because the MAR assumption cannot be verified from the observed data, the most transparent and straightforward approach is to impute missing observations under some MNAR scenarios and evaluate the robustness of inferences against MAR. Like the ANCOVA approach, we use the pattern mixture models for generating imputations under MNAR and define the extended model

$$Y_{ij} = \left(\beta_0 + \delta_0\right) + \left(\beta_1 + \delta_1\right)G_i + \left(\beta_2 + \delta_2\right)T_{ij} + \left(\beta_3 + \delta_3\right)G_i \times T_{ij} + G_{0i} + \lambda R_{ij}, \quad (5.16)$$

where the sensitivity parameters δ's and λ identify departure from the MAR assumption. Formulation of Equation 5.16 postulates an equal random component τ_0^2 across four patterns, as it might be unlikely to assume otherwise.

Following the principle of the pattern mixture model, three versions of the extended model (Equation 5.16) can be specified, one for each of the missing data patterns. In pattern 2, the pre-measurement scores are fully observed implying that $\delta_0 = \delta_1 = 0$ in this pattern (as there are no missing observations at baseline for males and females, respectively). The last two patterns (patterns 3 and 4) consist of less than 2% of data, wherein a total of five subjects had missing pre-measurement score (see Table 5.9). Because well-being at baseline is measured before the life event, it is less likely that these subjects have systematically different well-being scores (at baseline) had these missing observations been observed. We may therefore conclude that δ_0 and δ_1 can be assumed zero in patterns 3 and 4 as well. Note that this assumption implicitly implies that the missingness of pre-measurement is MCAR or MAR. We also make a further simplification by fixing $\lambda = 1$ as in the ANCOVA approach. Hence, the extended random-intercept model (Equation 5.16) simplifies to

$$Y_{ij} = \beta_0 + \beta_1 G_i + (\beta_2 + \delta_2) T_{ij} + (\beta_3 + \delta_3) G_i * T_{ij} + G_{0i} + R_{ij}, \qquad (5.17)$$

which can be applied to each of three missing data patterns. Specifically, Equation 5.17 is used to impute missing well-being scores at post measurement in patterns 2 and 4 after assigning plausible values to the sensitivity parameters δ_2 and δ_3. This model, however, simplifies to $Y_{ij} = \beta_0 + \beta_1 G_i + G_{0i} + R_{ij}$ for imputation of missing well-being measurements at baseline (as $T_{ij} = 0$ at baseline). Specifying plausible values for δ_2 and δ_3 is therefore the last step to complete the task of sensitivity analysis.

Baseline comparison of well-being scores, among those who remained and those who did not remain in the study after the life event, revealed that the latter group scored lower than the former at three months after the life event (i.e. at post measurement). This suggests that persons who scored or would have scored low at post measurement may be more likely to drop out after the life event. Moreover, a closer look at data shows that the older the person is, the more likely the person drops out after the life event. Also, it is expected that older persons would have poorer well-being after the life event. All in all, it can be argued that persons with missing post measurement scores perhaps could score low had these scores been observed.

Following the above arguments, we may assign negative values to δ_2 and δ_3, as these will lower the imputed values under MAR. We have also noticed that gender is not a predictor of missingness at post measurement, which, in turn, suggests that both sensitivity parameters are equal.

Table 5.10 shows the results of various analyses on the life-event study. The analysis model is the random-intercept model (Equation 5.15) with time, gender and their interaction as the fixed-effects factors. We have focussed on what would be the gender effect on well-being if all participants remained in the study and provided the well-being score three months after the life event.

The first row of Table 5.10 relates to the analysis of the observed data under the MAR assumption, showing a statistically significant difference at the 5% level between males and females. A more plausible assumption is that the data are MAR given time, gender, age and group. For this analysis, we used age and group as auxiliary variables in the imputation phase and subsequently fitted Equation 5.15 to each imputed dataset (m = 500 imputations). The pooled estimate of the gender effect decreases to −2.39, with a reduced p-value. The third analysis decreases the MAR imputed value (in the second analysis) by 0.10, followed by a gradual decline to 0.50 and 1. The gender effect estimate based on different values for the sensitivity parameters is gradually dropped as a result of more deviation from the MAR imputation. Nevertheless, the effect remained still significant at the 5% level, even for extremely large values $\delta_2 = \delta_3 = 1$ or $\delta_2 = 0$, $\delta_3 = 1$. We conclude that, under the MNAR mechanism represented by this particular model for sensitivity analysis, the inference is robust to MAR.

TABLE 5.10

Life-event Study: Estimates of the Gender Effect from Observed Data under MAR (row 1), from Multiple Imputation under MAR (row 2) and from MNAR with Different Values for Sensitivity Parameters δ_2 and δ_3

Mechanism	Estimate	SE	T-value	95% CI
MAR	−2.30	0.85	−2.71	(−3.98, −0.63)
Stronger MAR	−2.39	0.86	−2.79	(−4.07, −0.71)
MNAR				
$\delta_2 = -0.1, \delta_3 = -0.1$	−2.42	0.86	−2.82	(−4.09, −0.74)
$\delta_2 = -0.2, \delta_3 = -0.2$	−2.44	0.86	−2.86	(−4.12, −0.77)
$\delta_2 = -0.4, \delta_3 = -0.4$	−2.49	0.86	−2.92	(−4.17, −0.82)
$\delta_2 = -0.5, \delta_3 = -0.5$	−2.52	0.86	−2.95	(−4.20, −0.84)
$\delta_2 = -1.0, \delta_3 = -1.0$	−2.65	0.86	−3.09	(−4.33, −0.97)
$\delta_2 = 0.0, \delta_3 = -0.2$	−2.44	0.86	−2.85	(−4.12, −0.76)
$\delta_2 = 0.0, \delta_3 = -1.0$	−2.64	0.86	−3.08	(−4.13, −0.96)
$\delta_2 = -0.2, \delta_3 = 0.0$	−2.39	0.86	−2.80	(−4.07, −0.72)

MI analyses used m = 500 imputations. The analysis Equation 5.15 is used to estimate parameters after MI.

5.4 Assignments

5.4.1 Assignment

Consider the analysis of the Well-being study as described in Section 5.2. Why is a random-intercept model with a diagonal R matrix the same as the unstructured marginal model?

5.4.2 Assignment

Consider the 'Beating the Blues' study (use the SPSS system files 'Ancova BTB.sav' and 'LongBTB.sav'). See section 'short description of research and simulation study' and Section 5.2 for a detailed description of the study. Consider the baseline measurement and the repeated measurement after 8 months. Use as significant level $\alpha = 0.1$.

a. Obtain descriptive statistics like number of observations, average and standard deviation per treatment group and discuss the results.
b. Analyse the treatment effect after 8 months, using the ANCOVA and gain-score approaches and discuss the results (assume that the missing data mechanism is MAR).
c. Which approach do you prefer and why?

5.4.3 Assignment

Consider the Life-event study (use the SPSS system files 'Lifeancova.sav 'and Lifesubset.sav'). See section 'short description of research and simulation study' and Section 5.2 for a detailed description of the study. Consider the pre-measurement and the repeated measurement after one year. Use as significant level $\alpha = 0.05$.

  a. Obtain descriptive statistics like number of observations, average and standard deviation per treatment group and discuss the results.
  b. Analyse the life-event effect after one year, using the ANCOVA and gain-score approaches and discuss the results.
  c. Which approach do you prefer and why?

5.4.4 Assignment

In the 'Beating the Blues' study, consider the measurements of baseline and study end (i.e. 12 months) as pre and post measurements. (use the SPSS system file 'BTB_wide.sav').

  a. Investigate whether history of depression ('duration' variable in the file) is associated with missingness at the end of the study. Hint: use logistic regression with a missing indicator of the post measurement as the dependent variable.
  b. Fit an ANCOVA model evaluating the effect of treatment on post measurement adjusted for baseline measurement and history of depression.
  c. Is there a difference between the treatment effect estimate in part b and Table 5.7? Motivate your answer.

5.4.5 Assignment

In the 'Beating the Blues' study, consider the measurements of baseline and study end (i.e. 12 months) as pre and post measurements. (use the SPSS system file 'BTB_wide.sav').

  a. Use the history of depression as an auxiliary variable for the imputation purpose only and subsequently fit Equation 5.12 to the imputed data (choose a very large number of imputations, e.g., 5000).
  b. Compare the estimated treatment effect with part b of assignment 5.4.4.
  Perform a sensitivity analysis on the imputed data in part a by setting $\delta_0 = 0, \delta_1 = 1.$, and $\delta_0 = 0, \delta_1 = 0.5$

5.4.6 Assignment

Consider the first two measurements in the Life-event study (i.e. baseline and 3 months after the life event). Use the SPSS system file 'Lifeancova.sav' file.

a. Obtain the patterns of missing data. How many subjects have complete data?
b. Calculate the gain score for subjects and compare the well-being score between males and females using the standard two independent samples t-test.
c. Transform data to the long format (see Chapter 4) and fit the random-intercept model (Equation 5.15) to the incomplete data.
d. Compare the results of part b and c.

5.4.7 Assignment

Given an RCT and the ANCOVA model (Equation 5.2).

a. Show that from equations 5.3 and 5.4, it follows that
$E(\hat{\beta}_2(\bar{Y}_{1,G=1} - \bar{Y}_{1,G=0})) = E(\bar{Y}^*_{2,G=1} - \bar{Y}_{2,G=0})$.
b. Argue that in the absence of treatment
$E(\bar{Y}^*_{2,G=1} - \bar{Y}_{2,G=0}) = E(\bar{Y}_{1,G=1} - \bar{Y}_{1,G=0}) = 0$.
c. Show that the uncorrected difference in post measurement $\hat{\beta}_1 = \bar{Y}_{2,G=1} - \bar{Y}_{2,G=0}$ is an unbiased estimator of the true treatment effect, that is show that $E(\hat{\beta}_1) = E(\bar{Y}^*_{2,G=1} - \bar{Y}_{2,G=1})$.
d. Show that the estimator of the gain-score approach in Equation 5.5 $\hat{\beta}_1 = \bar{Y}_{2,G=1} - \bar{Y}_{1,G=1} - (\bar{Y}_{2,G=0} - \bar{Y}_{1,G=0})$ is unbiased, that is show that $E(\bar{Y}^*_{2,G=1} - \bar{Y}_{1,G=1}) = E(\bar{Y}_{2,G=0} - \bar{Y}_{1,G=0}) = 0$.
e. Show that Equation 5.8 holds with
$E(\bar{Y}^*_{2,G=1} - \bar{Y}_{1,G=1}) = E(\bar{Y}_{2,G=0} - \bar{Y}_{1,G=0}) = 0$. (Use the property: $E(X) - E(Y) = E(X-Y)$)

6

Analysis of Longitudinal Life-Event Studies

6.1 Best Practice for the Analysis of a Longitudinal Life-Event Study

In this section, we demonstrate how to analyse the Well-being study wherein the data are collected at three time points. The main goal in this chapter is to study how the difference between widow(er)s and caregivers changes over time due to the life event and whether this difference recovered towards the baseline situation. Thus, there are three pairwise comparisons to be made: the change in difference from baseline to three months after the life event (short-time effect), the change in difference from baseline to 12 months after the life event (long-time effect) and the change from 3 to 12 months after the life event.

We start with the observed profiles of average well-being scores of care-givers and widow(er)s. Because the profile might be different among males and females, Figure 6.1 displays the averages of well-being in caregivers (solid line) versus widowers (dashed line) for males in the left panel and for females in the right panel. From these profiles, the following observations can be made:

- The difference in average well-being between caregivers and widow(er)s increases after 3 months. This effect is similar for males and females.

- There is recovery in the well-being status 12 months after the life event, in the sense that the difference in average well-being after 12 months changes towards the baseline situation. This effect is similar for males and for females.

6.1.1 Choosing between the Gain-Score and ANCOVA Approach

Suppose we want to evaluate the above-mentioned observations. If partici-pants are pre-existing groups, for example men versus women, an analysis

DOI: 10.1201/9781003121381-6

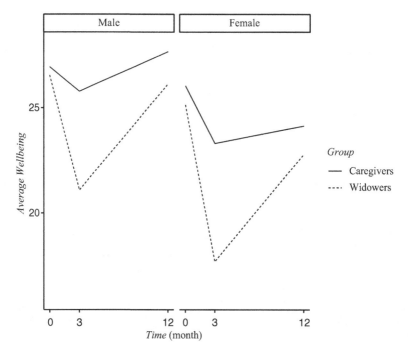

FIGURE 6.1
Average profiles of *Well-being* for caregivers and widow(er)s separately for males and females in the Well-being study.

based on the gain-score approach may be a better choice than the ANCOVA approach (see Chapter 5). However, in this particular study, the groups can only be distinguished after the life event. It is thus unclear whether the widow(er) and caregivers are coming from two different groups before the life event; for example, there may already be lifestyle or genetic differences between the two groups.

To be on the safe side, it is probably wise to analyse the data with both gain-score analysis and ANCOVA, so that the reliability of the results and conclusions from the two approaches can be compared. If both analyses, on the one hand, lead to similar results and interpretation, approach selection is a matter of taste. If both analyses, on the other hand, lead to a different interpretation, presentation of both results is advisable, especially in the situation of conflicting conclusions. Although this may seem odd at first, we should realise that the information regarding counterfactual profiles (as explained in Chapter 5) is not available. Hence, *a priori* preference cannot be made between both approaches.

In Sections 6.1.2 and 6.1.3, we discuss the gain-score and ANCOVA approach together with their corresponding sensitivity analyses.

6.1.2 Gain-Score Approach of the Well-Being Study

We will first investigate the issue of missing observations in the Well-being study, as longitudinal studies often have missing observations because subjects drop out permanently or skip visits intermittently. Table 6.1 presents all patterns of missing data for this study. Apart from the well-being measurements, the other variables (i.e. *Age*, *Gender* and *Group*) are complete. Two participants (pattern 6) did not provide the well-being score at all and thus will be excluded from the analysis. Three other participants had intermittent patterns of missing data (patterns 4 and 5), while the rest showed a monotone pattern of missing data (patterns 1–3).

Ignoring participants in patterns 4 and 5 for the moment (less than 1%), a monotone pattern for missing data can be concluded. This implies that the rate of missing data increases with *Time* so that the missingness depends on *Time*. Therefore, the MCAR assumption may be unlikely. We further investigate the plausibility of MCAR by identifying key predictors of missingness by a logistic regression of missingness in well-being at 3 and 12 months after the life event on available baseline variables (Tables 6.2a and 6.2b for missingness at 3 and 12 months after the life event, respectively). From both

TABLE 6.1

Well-being Study: the Patterns of Missing Data

Pattern	Baseline	After Three Months	After One Year	Number
1	O	O	O	184
2	O	O	M	17
3	O	M	M	63
4	M	O	O	2
5	M	O	M	1
6	M	M	M	2
Observed	98%	76%	69%	
	264	204	186	

TABLE 6.2A

Well-Being Study: Results of a Logistic Regression with Missingness 3 Months After the Live Event as the Dependent Variable

	B	S.E.	Wald	df	p-value
Age	0.025	0.02	1.15	1	0.284
Group	−11.17	3.94	8.04	1	0.005
Age × Group	0.14	0.05	7.43	1	0.006
Constant	−2.74	1.69	2.62	1	0.106

TABLE 6.2B

Well-Being Study: Results of a Logistic Regression with Missingness 12 Months After the Live Event as the Dependent Variable

	B	S.E.	Wald	df	p-value
Age	0.035	0.02	2.37	1	0.124
Group	−10.40	3.64	8.19	1	0.004
Age × Group	0.13	0.05	7.28	1	0.007
Constant	−3.00	1.60	3.53	1	0.060

models, it appears that the interaction between *Age* and *Group* is significant at the 5% level showing larger odds of withdrawal among older widow(re)s compared with their counterparts. This additionally confirms implausibility of the MCAR assumption in this study. Consequently, an analysis restricted to complete cases (i.e. an analysis based on the first pattern only) will likely produce biased estimates.

We thus start with the MAR assumption and specify the fixed part of the analysis model by *Time, Gender, Group, Age,* and all their first-order interaction terms. Later, we evaluate the robustness of inference with respect to the MAR assumption by exploring the reasons for missing data and subsequently perform sensitivity analysis using MI.

According to the basic guidelines in Section 4.1, we have to determine the best variance-covariance structure of the errors in the analysis model first. As mentioned earlier in Section 3.3.1, an unstructured (UN) variance-covariance structure is a safe choice when there are three time points. The interested readers, however, may follow the basic guidelines (explained in Chapter 4) to determine the best variance-covariance structure (with the smallest number of parameters and without a significant loss of information).

Using the UN variance-covariance structure, we apply the backward procedure to determine the fixed part the model by selecting variables for which their corresponding regression estimates are significantly different from zero (at the 5% level). It appears that the final model is

$$Well\,b_{ij} = \beta_0 + \beta_1 Group_i + \beta_2 Gender_i + \beta_3 T_{3,i} + \beta_4 T_{12,i}$$
$$+ \beta_5 Group_i \times T_{3,i} + \beta_6 Group_i \times T_{12,i} + \beta_7 Gender_i \times T_{3,i} \qquad (6.1)$$
$$+ \beta_8 Gender_i \times T_{12,i} + \beta_9 Age_i + V_{ij}.$$

In Equation 6.1, the time points are treated as a discrete variable since we are interested in the short and long-time effects (see also Chapters 2 and 6). Moreover, the observed profiles in Figure 6.1 do not suggest that straight lines can appropriately describe the change over time. Thus, two dummy variables T_3 and T_{12} must be defined such that $T_{3,i} = 1$ ($T_{12,i} = 1$) if 3 (12) months after the life event and $T_{3,i} = 0$ ($T_{12,i} = 0$) otherwise for $i = 1,...,n$. Note that the baseline measurement arbitrarily serves as the reference time point in Equation 6.1. Another time point could be chosen as the reference time point in order to compare, for example the difference between widow(er)s and caregivers 3 months and 1 year after the life event. Because the interaction term *Group × Gender* is not statistically significant (at the 5% level), it can be concluded that the difference between caregivers and widow(er)s is the same among males and females. It appears that this applies for each time point (due to the nonsignificant second order interaction *Group × Gender × Time* ; not presented here). This confirms what we see in Figure 6.1, as both plots (left and right) suggest a similar pattern across each gender group. A similar interpretation can be given for the interaction terms *Group × Age* and *Gender × Age*.

To interpret the regression parameters of Equation 6.1, the best way is to express them in terms of the coding scheme for the variables

$$\beta_1 = \mu_{0,wido(we)r} - \mu_{0,carer}$$
$$\beta_2 = \mu_{0,female} - \mu_{0,male}$$
$$\beta_5 = \left(\mu_{3,widow(er)} - \mu_{3,carer}\right) - \left(\mu_{0,widow(er)} - \mu_{0,carer}\right)$$
$$\beta_6 = \left(\mu_{12,widow(er)} - \mu_{12,carer}\right) - \left(\mu_{0,widow(er)} - \mu_{0,carer}\right) \tag{6.2}$$
$$\beta_7 = \left(\mu_{3,female} - \mu_{3,male}\right) - \left(\mu_{0,female} - \mu_{0,male}\right)$$
$$\beta_8 = \left(\mu_{12,female} - \mu_{12,male}\right) - \left(\mu_{0,female} - \mu_{0,male}\right),$$

where $\mu_{t,gender}$, $t = 0, 3, 12$ is the population average of *Well-being* at baseline, 3 and 12 months after the life event per category of *Gender*, respectively. Likewise, $\mu_{t,group}$ is the population average of *Well-being* at different time points per category of *Group*.

Note that the regression model with Equation 6.1 is conditional on *Gender*, *Age* and *Group*, which implies the regression parameters are interpreted by keeping the value of these variables constant. For instance, β_1 is the average difference between the well-being of widow(er)s and caregivers at baseline in the population when *Gender* and *Age* are fixed (see Chapter 2 for an interpretation of the regression parameters). For the ease of presentation, we do not mention the conditioning on *Group, Gender* and *Age* in further discussion.

The regression parameter β_2 is the difference in population average of *Well-being* of males and females at baseline. The regression parameters β_5 and β_6 represent the difference in population average of *Well-being* between widow(er)s and caregivers 3 and 12 months after the life event, respectively, which is fully corrected for the difference at baseline (according to the gain-score approach). The regression parameters β_7 and β_8 can be similarly interpreted with respect to the difference in population average of *Well-being* between males and females, fully corrected for baseline difference in average Well-being. Table 6.3 shows the parameter estimates of Equation 6.1.

TABLE 6.3

Well-being Study: the REML Estimates of the Fixed Effect Parameters using the Gain-score Approach

Parameter	Estimate	Std. Error	df	t-value	p-value	95% Confidence Interval	
						Lower Bound	Upper Bound
Intercept $\hat{\beta}_0$	33.63	2.73	260.65	12.34	0.000	28.26	39.00
Group $\hat{\beta}_1$	−0.61	0.56	262.64	−1.10	0.275	−1.71	0.49
Gender $\hat{\beta}_2$	−1.52	0.61	268.02	−2.49	0.013	−2.72	−0.32
$T_3\,\hat{\beta}_3$	−1.16	0.73	210.07	−1.60	0.112	−2.59	0.27
$T_{12}\,\hat{\beta}_4$	0.48	0.76	202.58	0.64	0.524	−1.01	1.98
$T_3 \times Group\,\hat{\beta}_5$	−4.60	0.77	210.60	−5.94	0.000	−6.13	−3.08
$T_{12} \times Group\,\hat{\beta}_6$	−0.97	0.80	198.76	−1.22	0.226	−2.54	0.60
$T_3 \times Gender\,\hat{\beta}_7$	−1.77	0.82	210.14	−2.16	0.032	−3.39	−0.15
$T_{12} \times Gender\,\hat{\beta}_8$	−2.27	0.85	201.15	−2.66	0.008	−3.95	−0.59
Age $\hat{\beta}_9$	−0.09	0.04	257.59	−2.45	0.015	−0.16	−0.02

Dependent Variable: *Well-being*.

Because several pairwise comparisons need to be performed, the overall type I error rate will become inflated. Therefore, a multiple comparison method is used to control for the increase of the overall type I error rate (see Chapter 4 for a discussion). Three pairwise comparisons are made: the change of the difference in average well-being between widow(er)s 3 months, 12 months after the life event and the change of the difference in average well-being from 3 to 12 months after the life event. Here, we use the Holm correction procedure and present the results in Table 6.4. Following the Holm procedure, it can be concluded that the difference in average well-being between widow(er)s and caregivers due to the life event after 3 months is significant $p_1 = 0.000$ (see also Table 6.3, where $\hat{\beta}_5 = -4.60$). However, the difference disappears after 12 months ($p_3 = 0.226$. See also Table 6.3, where $\hat{\beta}_6 = -0.97$). Note that to evaluate the change in group difference from 3 to 12 months after the life event, the coding scheme for *Time* must be changed (see Chapter 2 for an illustration). It turns out that the corresponding p-value for this latter test is zero ($p_2 = 0.000$) showing enough evidence to reject the hypothesis of no change in group difference from 3 to 12 months after the life event. This conclusion applies to both males and females, as there was no *Gender × Group* interaction (see also Figure 6.1 at the sample level). It should be noted that the presented correction methods for multiple comparisons are most suitable for uncorrelated hypotheses. As a result, these procedures are still conservative in longitudinal studies because hypotheses are usually correlated in repeated measurement designs. Afshartous and Wolf (2007) proposed a method for dependent hypotheses, which may provide a more accurate correction for multiple comparisons in longitudinal settings (based on the bootstrap, which goes beyond the scope of this book).

6.1.2.1 Sensitivity Analysis for Gain-Score Approach using Multiple Imputation

The analysis of well-being data assumed the MAR mechanism for the *Well-being* conditional on *Age, Gender, Group* and possibly their first-order interaction. As the observed data cannot be used to verify the MAR assumption, we

TABLE 6.4

Well-being Study: the Holm's Procedure to Compare Caregivers and Widow(er)s at Different Time Points

$H_0 : \Delta\mu_{cw,0} = \Delta\mu_{cw,3}$	$H_0 : \Delta\mu_{cw,3} = \Delta\mu_{cw,12}$	$H_0 : \Delta\mu_{cw,0} = \Delta\mu_{cw,12}$
$p_1 = 0.000^{\#}$	$p_2 = 0.000^{\#}$	$p_3 = 0.226$
$\alpha_1^* = 0.017$	$\alpha_2^* = 0.025$	$\alpha_3^* = 0.05$

Significant after the Holm's correction. $\Delta\mu_{cw,t}$ is the population average difference between caregivers and widow(er)s at time point t.

use multiple imputation for conducting sensitivity analysis to explore the robustness of inferences against violations of the MAR assumption.

The imputation process is done in the wide format using the MICE approach, that is missing observations are imputed on a variable-by-variable basis. To fulfil the inclusive strategy discussed in Chapter 4, the imputation models include *Age, Group, Gender* and *Age × Group*. The last interaction is added in the imputation models, as this was associated with the missingness at 3 and 12 months after the life event (see Tables 6.2a and b). The following set of imputation models is thus specified for the well-being before, 3 and 12 months after the life event:

$$
\begin{aligned}
Well\,b_{i0} &= \alpha_{00} + \alpha_{10}Group_i + \alpha_{20}Gender_i + \alpha_{30}Age_i + \alpha_{40}Age_i \\
&\quad \times Group_i + \alpha_{50}Well\,b_{i3} + \alpha_{60}Well\,b_{i12} + V_{i0} \\
Well\,b_{i3} &= \alpha_{01} + \alpha_{11}Group_i + \alpha_{21}Gender_i + \alpha_{31}Age_i + \alpha_{41}Age_i \\
&\quad \times Group_i + \alpha_{51}Well\,b_{i0} + \alpha_{61}Well\,b_{i12} + V_{i3} \\
Well\,b_{i12} &= \alpha_{02} + \alpha_{12}Group_i + \alpha_{22}Gender_i + \alpha_{32}Age_i + \alpha_{42}Age_i \\
&\quad \times Group_i + \alpha_{52}Well\,b_{i0} + \alpha_{62}Well\,b_{i3} + V_{i12},
\end{aligned}
\tag{6.3}
$$

where *Well* b_{i0}, *Well* b_{i3}, and *Well* b_{i12} are the well-being before, 3 and 12 months after the life event, respectively. As discussed in Chapter 4, the imputation Equations 6.3 imply an unstructured variance-covariance structure for the *Well-being* so that this structure is consistent with the variance-covariance structure in the analysis Equation 6.1 (which will be a UN variance-covariance structure), and so the inclusive strategy is not violated.

Using the imputation Equations 6.3, multiple imputation is carried out essentially under the MAR assumption. To explore how deviations from MAR may change the conclusions, we extend the imputation models 6.3 to accommodate some MNAR mechanisms. In the Well-being study, a plausible scenario is that participants with missing observations might have less well-being than what will be imputed under MAR. This implies that the imputed values obtained from Equations 6.3 should be lowered. This can be done for either caregivers or widow(er)s or even for both groups. Hence, we specify the imputation models under the specific MNAR mechanism by

$$
\begin{aligned}
Well\,b_{i0} &= \left(\alpha_{00} + \delta_{00}\right) + \left(\alpha_{10} + \delta_{10}\right)Group_i + \alpha_{20}Gender_i + \alpha_{30}Age_i + \alpha_{40}Age_i \\
&\quad \times Group_i + \alpha_{50}Well\,b_{i3} + \alpha_{60}Well\,b_{i12} + V_{i0} \\
Well\,b_{i3} &= \left(\alpha_{01} + \delta_{01}\right) + \left(\alpha_{11} + \delta_{11}\right)Group_i + \alpha_{21}Gender_i + \alpha_{31}Age_i + \alpha_{41}Age_i \\
&\quad \times Group_i + \alpha_{51}Well\,b_{i0} + \alpha_{61}Well\,b_{i12} + V_{i3} \\
Well\,b_{i12} &= \left(\alpha_{02} + \delta_{02}\right) + \left(\alpha_{12} + \delta_{12}\right)Group_i + \alpha_{22}Gender_i + \alpha_{32}Age_i + \alpha_{42}Age_i \\
&\quad \times Group_i + \alpha_{52}\alpha_{52}Well\,b_{i0} + \alpha_{62}Well\,b_{i3} + V_{i12},
\end{aligned}
\tag{6.4}
$$

here the δ's are sensitivity parameters. Each δ controls how much an imputed value differs from the corresponding imputation under MAR within each group and time point. For example, $\delta_{02} < 0$ and $\delta_{12} = 0$ imply that the well-being at 12 months after the life event is lower than predicted under MAR in both caregiver and widow(er) groups, while $\delta_{02} = 0$ and $\delta_{12} < 0$ mean a lower score only in the widow(er) group. Although we do not introduce additional sensitivity parameters in the imputation Equations 6.4, the formulation of MNAR models is general enough to accommodate any other scenarios that might seem plausible. This means sensitivity parameters could be specified for other independent variables such as *Gender* or *Age*.

The imputation process under MNAR is further simplified by allowing $\delta_{00} = 0$ and $\delta_{10} = 0$ because we do not anticipate that participants with missing observations at baseline (i.e. before the life event) will have a different well-being status than predicted under MAR. Moreover, there were only three participants (less than 1%) with missing observations at baseline so that the investigation of deviations from MAR at baseline seems trivial.

We finally explore six plausible scenarios summarised in Table 6.5a and Table 6.5b. Table 6.5a displays the estimate of the regression parameters and their corresponding standard errors for the selected δ values. The notation of the regression parameters follows that of Equation 6.4. Table 6.5b displays the p-values for selected δ values, following the Holm procedure.

The first scenario leads to imputations under MAR because the imputation Equations 3.6 are equivalent to the imputation Equations 3.5 when all sensitivity parameters are zero. Similar to Chapter 5, the sensitivity parameter values for the other scenarios are selected based on the raw average differences in well-being between those who remained in the study and those who dropped out. For example, we compared caregivers who had missing observations at 3 months after the life event and those who did not. The average of well-being at baseline in these two groups differed about one unit. The same

TABLE 6.5A

Well-being Study: the REML Estimates of the Fixed Effects Parameters (with Standard Errors) for Selected Sensitivity Parameters using the Gain-Score Approach

Scenario	Sensitivity Parameter	$\hat{\beta}_5$	$\hat{\beta}_6$	$\hat{\beta}_6 - \hat{\beta}_5$
	$(\delta_{01}, \delta_{11}, \delta_{02}, \delta_{12})$			
1	$(0, 0, 0, 0)$	$-4.48\ (0.79)$	$-0.95\ (0.83)$	$3.53\ (0.68)$
2	$(0, -1, 0, -2)$	$-5.04\ (0.80)$	$-1.74\ (0.84)$	$3.30\ (0.68)$
3	$(-1, -1, 0, -1)$	$-4.76\ (0.80)$	$-1.35\ (0.83)$	$3.42\ (0.68)$
4	$(-1, -1, 0, -2)$	$-4.90\ (0.80)$	$-1.66\ (0.84)$	$3.24\ (0.68)$
5	$(-1, -1, 0, -4)$	$-5.19\ (0.81)$	$-2.29\ (0.87)$	$2.89\ (0.69)$
6	$(-2, -2, -1, -2)$	$-4.98\ (0.84)$	$-1.58\ (0.87)$	$3.40\ (0.68)$

TABLE 6.5B

Well-being Study: the Holm's Procedure to Compare Caregivers and Widow(er)s at Different Time Points for Selected Sensitivity Parameters using the Gain-Score Approach

Scenario	Sensitivity Parameter	$H_0: \Delta\mu_{cw,0} = \Delta\mu_{cw,3}$	$H_0: \Delta\mu_{cw,3} = \Delta\mu_{cw,12}$	$H_0: \Delta\mu_{cw,0} = \Delta\mu_{cw,12}$
	$(\delta_{01}, \delta_{11}, \delta_{02}, \delta_{12})$			
1	$(0, 0, 0, 0)$	<0.001*	<0.001*	0.206
2	$(0, -1, 0, -2)$	<0.001*	<0.001*	0.046
3	$(-1, -1, 0, -1)$	<0.001*	<0.001*	0.108
4	$(-1, -1, 0, -2)$	<0.001*	<0.001*	0.057
5	$(-1, -1, 0, -4)$	<0.001*	<0.001*	0.012*
6	$(-2, -2, -1, -2)$	<0.001*	<0.001*	0.078
	adjusted significance level	$\alpha_1^* = 0.0167$	$\alpha_2^* = 0.025$	$\alpha_3^* = 0.05$

* significant after the Holm's correction. $\Delta\mu_{cw,t}$ is the population average difference between caregivers and widow(er)s at time point t.

pattern is observed for the widow(er)s as well. Specifically, the average of well-being at baseline for widow(er)s who had a missing well-being score at 3 months after the life event was 24.8, while it was 25.7 for those who did not have missing data at 3 months after the life event. Hence, the imputed values at 3 months after the life event are lowered by one unit for both caregivers and widow(er)s in scenarios 3–5. Scenario 6 even assumes a bigger drop by setting $\delta_{01} = \delta_{11} = -2$. In contrast, it is assumed that widow(er)s have only poorer well-being in scenario 2 since $\delta_{01} = 0$ but $\delta_{11} = -1$.

To assign plausible values for δ_{02} and δ_{12}, we compared two groups: those who completed the study (i.e. had no missing observations) and those who dropped out at 12 months after the life event. Among caregivers, the average difference in well-being between these two groups was approximately zero (at baseline) and one (at 3 months after the life event), and therefore, δ_{02} varied between zero and one. In contrast, these differences were, respectively, around zero and four in the widow(er) group. Consequently, δ_{12} took values in the range between zero and four.

Multiple imputation was carried out with 50 imputations (k =50) per scenario, and the MICE algorithm was iterated 10 times per imputation.

The results of multiple imputation under MAR (scenario 1) agree with the mixed model results in Table 6.4. Imputing under MNAR (scenarios 2–6), we see the parameter estimates do not change substantially across scenarios, albeit to some extent in $\hat{\beta}_6$, but the overall conclusion holds as before (i.e. group effect due to the life event after 3 months is significant, but this disappears 12 months after the life event). Nevertheless, the sensitivity analysis reveals that there might not be a full recovery in group effect after 12 months because the null hypothesis $H_0 : \Delta\mu_{cw,0} = \Delta\mu_{cw,12}$ (column 5 in Table 6.5b) shows much reduced p-values under the MNAR scenarios as compared to the MAR in scenario 5. We conclude that, given the MNAR models represented in this example, the inference regarding the recovery of participants 12 months after the life event is not robust against the MAR assumption.

6.1.3 ANCOVA Approach of the Well-Being Study

Now we compare the abovementioned results with the results of the ANCOVA approach. Using the unstructured variance-covariance structure, the backward procedure is applied to select variables for which their corresponding regression estimates are significantly different from zero (at the 5% level). Starting with the full ANCOVA-approach model including all baseline independent variables and their first-order interaction terms, it appears that the final model is

$$Well\,b_{ij} = \beta_0 + \beta_1\,Group_i + \beta_2\,Gender_i + \beta_3\,T_i + \beta_4\,Well\,b_{0,i} \qquad (6.5)$$
$$+\,\beta_5\,Group_i \times T_i + V_{ij},$$

where $Well\,b_{0,i}$ is the well-being of subject i, $i = 1, \cdots, n$, at baseline, T_i is the time indicator coded as zero for 3 months after the life event and one for 12 months after the life event, and $Group_i$ and $Gender_i$ are defined as before (see Section 5.2.2). Here, V_{ij}, $j = 3, 12$, follows an unstructured (2×2) variance-covariance structure. Note that Equation 6.5 does not have any baseline dependent variable $Well\,b_{0,i}$ in the left side of the equation. Moreover, there is no $Gender \times Time$ interaction in the final ANCOVA model (see assignment 6.4.1.d), so that the model with Equation 6.5 is different from the model with Equation 6.1.

Like the gain-score approach, regression coefficients in Equation 6.5 are interpreted based on the codes for independent variables. Specifically,

$$\beta_1 = \left(\mu_{3,widow(er)} - \mu_{3,carer}\right) - \beta_4\left(\mu_{0,widow(er)} - \mu_{0,carer}\right)$$
$$\beta_2 = \left(\mu_{3,female} - \mu_{3,male}\right) - \beta_4\left(\mu_{0,female} - \mu_{0,male}\right) \qquad (6.6)$$
$$\beta_5 = \left(\mu_{12,widow(er)} - \mu_{12,carer}\right) - \left(\mu_{3,widow(er)} - \mu_{3,carer}\right)$$

Note that, in this case, the population averages $\mu_{t,group}$ and $\mu_{t,gender}$ are not expressed conditional on *Age* because the contribution of *Age* is not statistically significant and therefore not included in equation 6.5. The regression parameter β_1 in Equation 6.5 can be interpreted as the difference in population average of *Well-being* between widow(er)s and caregivers 3 months after the life event but corrected for a fraction of the difference at baseline. The regression parameter β_2 can be interpreted likewise. Finally, the regression parameter β_5 compares the difference in population average of *Well-being* between widow(er)s and caregivers 12 months after the life event and the difference 3 months after the life event. Also, the difference in population average of *Well-being* between males and females 3 months after the life event is the same as the difference after 12 months because there is no interaction between *Gender* and *Time* in Equation 6.5 (see also Figure 6.1). Table 6.6 shows the estimates of the regression parameters in Equation 6.5.

The reader may check that both approaches agree and lead to similar conclusions (see also assignment 6.4.1). However, this is not always the case as discussed in Section 6.1.1.

Which model should be presented in this case is a matter of convenience. Hence, the model that is easier to interpret given the research question can be presented. For instance, we could opt for the ANCOVA approach because the model specification looks simpler. However, the model specification of the gain-score approach describes the difference in *Well-being* that is adjusted for the baseline differences.

6.1.3.1 Sensitivity Analysis for ANCOVA Approach Using Multiple Imputation

Similar to the gain-score approach, the ANCOVA analysis makes the MAR assumption for missing observations in the Well-being study. We further investigate the robustness of the results with respect to the posited MAR mechanism by considering some MNAR scenarios. For this purpose, the MNAR imputed values in Section 6.1.2 are used as input to fit the ANCOVA model (Equation 6.5) to the imputed datasets after which the estimates are

TABLE 6.6

Well-Being Study: The REML Estimates of the Fixed Effect Parameters Using the ANCOVA Approach

Parameter	Estimate	Std. Error	df	t-value	p-value	95% Confidence Interval Lower Bound	Upper Bound
Intercept $\hat{\beta}_0$	12.20	2.17	195.17	5.62	0.000	7.92	16.48
Group $\hat{\beta}_1$	−4.93	0.73	198.97	−6.74	0.000	−6.37	−3.49
Gender $\hat{\beta}_2$	−2.48	0.70	196.69	−3.55	0.000	−3.86	−1.10
$T\,\hat{\beta}_3$	1.38	0.47	187.20	2.92	0.004	0.45	2.31
$Y_0\,\hat{\beta}_4$	0.51	0.08	190.01	6.75	0.000	0.36	0.66
Group × Time $\hat{\beta}_5$	3.52	0.70	185.60	5.06	0.000	2.15	4.89

Dependent Variable: *Well-being*.

pooled to obtain a single inference. Table 6.7a reports the estimates of regression coefficients in the ANCOVA model Equation 6.5 showing some degree of agreement with the previous results. Applying the Holm's procedure to the results from the ANCOVA approach indicates that recovery after 12 months is less evident (see Table 6.7b).

TABLE 6.7A

Well-Being Study: The REML Estimates of the Fixed Effects Parameters (with Standard Errors) for Selected Sensitivity Parameters using the ANCOVA Approach

Scenario	Sensitivity Parameter	$\hat{\beta}_1$	$\hat{\beta}_5$	$\hat{\beta}_5 + \hat{\beta}_1$
	$(\delta_{01}, \delta_{11}, \delta_{02}, \delta_{12})$			
1	$(0, 0, 0, 0)$	-4.82 (0.74)	3.48 (0.67)	-1.35 (0.78)
2	$(0, -1, 0, -2)$	-5.38 (0.74)	3.26 (0.67)	-2.13 (0.80)
3	$(-1, -1, 0, -1)$	-5.09 (0.75)	3.37 (0.67)	-1.72 (0.79)
4	$(-1, -1, 0, -2)$	-5.23 (0.75)	3.20 (0.67)	-2.04 (0.80)
5	$(-1, -1, 0, -4)$	-5.52 (0.76)	2.85 (0.68)	-2.66 (0.83)
6	$(-2, -2, -1, -2)$	-5.29 (0.80)	3.36 (0.68)	-1.93 (0.84)

TABLE 6.7B

Well-Being Study: The Holm's Procedure to Compare Caregivers and Widow(er) s at Different Time Points for Selected Sensitivity Parameters Using the ANCOVA Approach

Scenario	Sensitivity Parameter	$H_0 : \Delta\mu_{cw,0} = \Delta\mu_{cw,3}$	$H_0 : \Delta\mu_{cw,3} = \Delta\mu_{cw,12}$	$H_0 : \Delta\mu_{cw,0} = \Delta\mu_{cw,12}$
	$(\delta_{01}, \delta_{11}, \delta_{02}, \delta_{12})$			
1	$(0, 0, 0, 0)$	<0.001*	<0.001*	0.091
2	$(0, -1, 0, -2)$	<0.001*	<0.001*	0.012*
3	$(-1, -1, 0, -1)$	<0.001*	<0.001*	0.038*
4	$(-1, -1, 0, -2)$	<0.001*	<0.001*	0.016*
5	$(-1, -1, 0, -4)$	<0.001*	<0.001*	0.003*
6	$(-2, -2, -1, -2)$	<0.001*	<0.001*	0.028*
	adjusted significance level	$\alpha_1^* = 0.0167$	$\alpha_2^* = 0.025$	$\alpha_3^* = 0.050$

* Significant after the Holm's correction. $\Delta\mu_{cw,t}$ is the population average difference between caregivers and widow(er)s at time point t.

6.2 Extension to More Than Three Time Points and Groups

We have seen earlier in this chapter how the regression parameters in longitudinal designs with two groups and three time points can be interpreted. When longitudinal designs consist of more than two groups and/or three time points, the situation in principle is not different but may become more complex. Sections 6.2.1. and 6.2.2 discuss an extension of the gain-score and ANCOVA model to more than two groups and/or three time points, respectively.

6.2.1 Interpretation of the Regression Parameters of a Gain-Score Model

Consider a hypothetical study where there exist G groups, and participants are followed over T time points. Furthermore, suppose that the regression model consists of the main effects for *Group* and *Time* together with their interaction. Ignoring the index of subject i for simplicity, the regression model can thus be specified, for example for $G = 3$ and $T = 3$, as follows

$$Y_t = \beta_0 + \beta_1\, Gr_1 + \beta_2\, Gr_2 + \beta_3\, T_1 + \beta_4\, T_2 + \beta_5\, Gr_1 \times T_1$$
$$+ \beta_6\, Gr_1 \times T_2 + \beta_7\, Gr_2 \times T_1 + \beta_8\, Gr_2 \times T_2 + V_t,$$

(6.7)

with Y_t as the continuous, quantitative dependent variable at time point $t = 0, 1, 2^1$. In Equation 6.7, Gr_1 and Gr_2 as well as T_1 and T_2 are dummy variables representing three groups and three time points, respectively. Note that one group serves as the reference group (here the first group, i.e. Gr_0). Likewise, one time point (here the first time point) is the reference time point. Their corresponding dummy variables are evidently not included in the model due to being redundant. Potential confounders can also be added to the model as additional independent variables (in the right side of Equation 6.7). In that case, the estimates of the regression parameters must be interpreted as adjusted for the confounders. Finally, the regression parameters of Equation 6.7 are interpreted as of Equation 6.1. For instance,

→ *The estimator $\hat{\beta}_1$ ($\hat{\beta}_2$) in Equation 6.7 is interpreted as the difference in average of the dependent variable between group 1 (group 2) and the reference group (i.e., group 0) at the reference time point $T = 0$.*

By choosing another group as reference (e.g. group 1), other pairwise comparisons (e.g. group 1 vs. group 2) can be obtained at the reference time point. It should be noted that the first time point (i.e. $T = 0$), which is the reference time point in the current formulation of Equation 6.7 is mostly the baseline time point. Now, if the purpose of the analysis is to estimate a life-event effect, the differences at a certain time point should be corrected for the differences at baseline. Hence, the above formulation matches exactly our need to address the required analysis. To clarify this point, suppose we are interested in the life-event effect after a certain time point (i.e. $T = 1, 2$) and let us define the baseline time point ($T = 0$) as the reference time point together with the reference group 0. Due to the nonrandomised nature of the study, the expected difference in the dependent variable between groups at time point $T = 1$ (and $T = 2$) should be at least corrected for the difference at baseline. Thus,

→ *The estimated parameter of the interaction effect of $Gr_g \times T_t$ equals to the difference in average of the* dependent variable *between group g and the reference group at time point t corrected for the same difference at baseline.*

For instance, $\hat{\beta}_5$ (in Equation 6.7) estimates the difference in average of the dependent variable between group $G = 1$ and the reference group $G = 0$ at time $T = 1$, which is fully corrected for the difference at baseline. As discussed in Section 6.1.1, this is exactly equal to the life-event effect according to the gain-score approach. The above interaction effect can also be interpreted as the change in group effect from baseline to time point t. By changing the reference time point, one can determine the change in group effect between two arbitrary time points.

6.2.2 Interpretation of the Regression Parameters of an ANCOVA-Model

Following the ANCOVA approach, we have seen that the baseline dependent variable should be specified as an independent variable in the marginal or random-effects models. Suppose, as an example, the number of groups is $G = 3$ with groups $g = 0, 1, 2$ and the number of time points is $T = 4$ with time points $t = 0, 1, 2, 3$. As before, $g = 0$ is the reference group and $t = 0$ is the baseline time point. The dependent variable Y_0 is the baseline variable. An ANCOVA model can then be expressed by (ignoring the index of subject i)

$$Y_t = \beta_0 + \beta_1 Gr_1 + \beta_2 Gr_2 + \beta_3 T_2 + \beta_4 T_3 + \beta_5 Y_0 + \beta_6 Gr_1 \\ \times T_2 + \beta_7 Gr_1 \times T_3 + \beta_8 Gr_2 \times T_2 + \beta_9 Gr_2 \times T_3 + V_t. \tag{6.8}$$

In Equation 6.8, dummy variables T_2 and T_3 are enough to specify the time points $t = 1, 2, 3$ as $T_2 = 0$ and $T_3 = 0$ implies the first time point after the life event (i.e., $T = 1$). This model resembles model 5.1, but with three groups instead of two. The parameters of this model have similar interpretation as in Chapter 5, for instance,

→ *The estimator $\hat{\beta}_1$ ($\hat{\beta}_2$) can be interpreted as the difference in average of the dependent variable between group 1 (group 2) and the reference group (i.e. $G = 0$) at time point $T = 1$, which is corrected for a fraction ($\hat{\beta}_4$) of the same difference at baseline (i.e. $T = 0$).*

By changing the reference group, other pairwise comparisons can be made, and the effect of interest can be determined at each time point. Finally, the change in group effect between two arbitrary time points (after the life event) can be obtained from the interaction effects with the proper reference time.

6.3 Assignments

6.3.1 Assignment

Life-event study (SPSS system file: Lifesubset.sav). (Nieboer et al. 1998)

a. Repeat the analysis of the well-being data. Follow the 'basic guidelines' mentioned in the course. Use both the gain score and the ANCOVA approach. For the gain score approach use, the SPSS system file 'Lifesubset.sav' file. For the ANCOVA approach use the SPSS system file 'Lifeancova.sav' file. Suppose that the main interest is the comparison between males and females. Basically, a similar analysis as described in this chapter is requested.

b. Is the change in difference between males and females from 3 to 12 months after the life event statistically significant?

c. What are your conclusions concerning the difference between male and female over time for both the gain score analysis and the ANCOVA analysis?

d. Explain why an interaction term between *Gender* and *Time* is specified in the gain-score analysis and not in the ANCOVA analysis?

e. Perform a sensitivity analysis to study the robustness of the MAR assumption.

f. Discuss the pros and cons of the multiple comparison procedures mentioned in this chapter.

g. Which model would you prefer in this case?

6.3.2 Assignment

Consider the Proximity study as discussed in Chapter 4. Perform the Holm-Bonferroni multiple comparison procedure to compare the average *Proximity* between successive time points of males and females, that is change in the average *Proximity* from the first (Occ_0) to the second occasion (Occ_1), from the second (Occ_1) to the third occasion (Occ_2), and from the third (Occ_2) to the fourth occasion (Occ_3).

6.3.3 Assignment

Discuss the similarity between the regression parameter of β_5 of Equation 6.6 and β_3 in Equation 2.9 with *Time* (*T*) instead of *School* and *Group* (*Gr*) instead of *Age*.

6.3.4 Assignment

Consider Equation 6.3. Interpret the regression parameters β_6, β_7, β_8, and β_9.

6.3.5 Assignment

Check the correctness of the interpretation of the regression parameters $\beta_1, \ldots, \beta_8$ in Equation 6.2.

6.3.6 Assignment

Check the correctness of the interpretation of the regression parameters β_1, β_2 and β_5 in Equation 6.6.

Note

1 Actually, $t = t_0, t_1, t_2$, because any sequence of 3 time points can be chosen and is not only restricted to 0, 1, 2. However, for ease of presentation, we choose the sequence as indicated.

7

Analysis of Longitudinal Experimental Studies

7.1 Best Practice for the Analysis of Beating the Blues Trial

Focusing on the Beating the Blues (BtB) study described in Chapter 5, we consider the following main research question:

> Is the effect of BtB in treating depression different from that of TAU during the course of the trial?

Due to randomisation, all baseline (prognostic) factors are not confounded with the treatment. Consequently, it is unnecessary to include these prognostic factors in the analysis to estimate the treatment effect unbiasedly. However, on the one hand, including prognostic factors do have an additional value because it will increase the precision of the treatment effect. On the other hand, these factors can still be used to indicate whether the randomisation was performed correctly. For instance, if the distributions of baseline variables considerably differ between treatment arms, it may indicate that the randomisation was not done properly. Moreover, researchers usually tend to test for significancy of the differences between treatment arms for the baseline variables. This, however, is a bad practice and should not be advocated in randomised trials according to the CONSORT2010 statement (Schulz et al., 2010). In fact, it has been discouraged by many during the last four decades (e.g. de Boer et al., 2015). An exploratory comparison should instead be done using a combination of clinical knowledge and common sense (Altman, 1985). For example, according to Altman (1985), one can consider relevant differences of averages (or percentages) and use a rule of thumb as a threshold (e.g. a 10% difference). Prognostic factors for which the difference in averages (or percentages) between treatment arms exceeds the threshold may be included in the analysis model.

If one or more baseline variables have missing observations, an imputation technique should be applied first because an analysis restricted to complete cases is inefficient. Although multiple imputation can be used, mean imputation (or substitution) of baseline variables is an excellent alternative irrespective of the missingness mechanism if baseline measurements are obtained before randomisation (Kayembe et al., 2022b). As opposed, missing observations in the dependent variable (e.g. *Depress* in BtB) pose another issue and should be treated differently as simple methods like mean imputation fail. We therefore need to carefully investigate the cause of missing data when the dependent variable has missing observations.

Table 7.1 shows the patterns of missing data in the BtB study that appear to be monotone (see Chapter 3.5.1) and limited to the dependent variable only. From this table, it can be seen that dropout (at each time point) is approximately similar in both treatment groups so that dropout seems to be unrelated to the treatment (denoted by *Treat*). However, a closer examination reveals that the missingness at time *t* may depend on the depression score (denoted by *Depress*) before *t* and possibly its interaction with *Treat* as briefly outlined below.

Figure 7.1 displays the mean of depression score by *Time* per treatment arm for available cases (solid line) and complete cases (dashed line), respectively.

The figure suggests that, in the BtB arm, there is a larger drop in average of depression score for complete cases compared with available cases. Consequently, complete case analysis could exaggerate the treatment effect because the average of depression score in the BtB arm for those who fully comply with the study protocol and have no missing observations, is smaller than that of patients who drop out from the study. To formally investigate the association between missingness and observed depression scores, however, a methodology will be needed that goes beyond the scope of this book. Hedeker and Gibbons (2006, pp. 291–294) showed that the time until dropout can be modelled using a discrete-time survival analysis (Singer and Willett, 2003) for this purpose.

We conclude that the MCAR mechanism seems implausible in this case, so that an analysis restricted to complete cases may result in biased estimates (see assignment 7.2.1).

Given the abovementioned considerations, we assume the MAR assumption and specify the mixed-effects regression Equation 7.1 to address the research question.

$$
\begin{aligned}
Depress_{ij} = {} & \beta_0 + \beta_1 Treat_i + \beta_2 Bdipre_i + \beta_3\, Time_{i2} \\
& + \beta_4\, Time_{i3} + \beta_5\, Time_{i5} + \beta_6\, Treat_i \times Time_{i2} \\
& + \beta_7\, Treat_i \times Time_{i3} + \beta_8\, Treat_i \times Time_{i5} + V_{ij},
\end{aligned}
\tag{7.1}
$$

In Equation 7.1, $Depress_{ij}$ is the depression score for subject i at time point $j = 2, 3, 5, 8$ months, $Treat_i$ represents the treatment arm (1 for BtB and 0 for TAU), $Bdipre_i$ is the baseline depression score, and V_{ij}'s are the correlated

TABLE 7.1

Beating the Blues Study: Patterns of Missing Data by Treatment Arm

Pattern	Baseline	Month2	Month3	Month5	Month8	Number (%)
TAU N = 48						
1	O	O	O	O	O	25 (52%)
2	O	O	O	O	M	4 (8%)
3	O	O	O	M	M	7 (15%)
4	O	O	M	M	M	9 (19%)
5	O	M	M	M	M	3 (6%)
Observed (%)	48 (100%)	45 (93%)	36 (75%)	29 (60%)	25 (52%)	

Pattern	Baseline	Month2	Month3	Month5	Month8	Number (%)
BtB N = 52						
1	O	O	O	O	O	27 (52%)
2	O	O	O	O	M	2 (4%)
3	O	O	O	M	M	8 (15%)
4	O	O	M	M	M	15 (29%)
5	O	M	M	M	M	0 (0%)
Observed (%)	52 (100%)	52 (100%)	37 (71%)	29 (56%)	27 (52%)	

Note: 'O' denotes observed and 'M' missing.

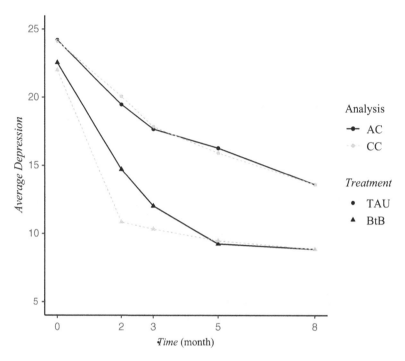

FIGURE 7.1
A comparison of average profiles of *Depression* for BtB and TAU groups between all available cases (AC) and complete cases (CC).

error terms. Because short and long-time effects are of interest (see also Chapters 2 and 6), *Time* is considered as a discrete variable and represented by the dummy variables $Time_{i2}$, $Time_{i3}$ and $Time_{i5}$ with 8 months as the reference. Although *Duration* is uncorrelated to *Treat*, it could also be included in the model to increase precision of the treatment effect. We, however, decided to omit this variable from Equation 7.1 to keep the presentation simple.

To investigate the fixed part of Equation 7.1, we first need to choose a proper covariance structure for the error terms. Following the basic guidelines discussed in Chapter 4, it appears that the compound symmetry model (i.e. the random-intercept model) fits the data best. It also appears that the *Time* × *Treat* interaction (represented by three interaction effects in Equation 7.1) is not significant at the 5% level (see assignment 7.2.2). As a result, the final model excludes all interaction terms in Equation 7.1. Table 7.2 shows the estimates of the regression coefficients in this final model. It can thus be concluded that, according to the ANCOVA approach, the difference in average of depression scores between both treatments does not significantly change over time after adjustment for the difference at baseline. The estimated treatment effect is nevertheless significant at the

TABLE 7.2

Beating the Blues: The REML Estimates of the Fixed-Effects Parameters Based on Equation 7.1 with a Compound Symmetry Variance-Covariance Structure and Without Interaction

Parameter	Estimate	Std. Error	df	t-value	p-value	95% Confidence Interval	
						Lower Bound	Upper Bound
Intercept	0.01	2.27	114.39	0.006	0.995	−4.48	4.51
Treat	−3.26	1.63	95.18	−2.006	0.048	−6.49	−0.03
Bdipre	0.62	0.08	97.90	8.094	0.000	0.46	0.77
$Time_2$	4.39	0.93	197.95	4.729	0.000	2.56	6.22
$Time_3$	2.96	0.94	188.16	3.136	0.002	1.10	4.82
$Time_5$	1.73	0.97	184.63	1.778	0.077	−0.189	3.64

Dependent Variable: *Depression* score.

5% level (p-value = 0.048), and the estimated average of depression score in the BtB group is 3.26 units smaller than in the TAU group.

Note that the abovementioned analysis is valid under the MAR assumption, that is the estimated treatment effect is unbiased if this assumption is met. The MAR assumption, however, is untestable (see Chapter 3), and so, the MNAR situation cannot be ruled out since it is still possible that the missingness depends on unobserved depression scores. It is conceivable that, for example, patients with missing values in the BtB arm might be those with high depression scores. Consequently, the estimate of the treatment effect would be lowered had these scores been observed.

We remind the reader that the missingness at time point t depends on the interaction between *Treat* and observed *Depress* before t. It suggests that, in the BtB arm, the higher the observed depression score of the patient, the more likely the patient to drop out. Hence, we investigate situations in which it is presumed that patients with a missing depression score in the BtB arm have a higher score than predicted under MAR.

Similar to Chapter 6, we develop imputation models for depression score in the wide format using the MICE approach, that is, missing depression scores at each time point are imputed conditional on the depression scores at other time points plus *Bdipre* and *Treat*. We also add *Duration* to the imputation model, as it is associated with the missingness. Because it is expected that patients in the BtB arm could have higher depression scores than predicted under MAR, we consider three plausible scenarios for obtaining imputations under MNAR.

Constant change: Imputed values at each time point are increased by a constant factor δ.

1. *Increasing change*: Imputed values at time point k are increased by $k \times \delta$ with $k = 0, 1, 2, 3$ at month 2, 3, 5 and 8 (at month 2, $k = 0$ as there are no missing observations in the BtB arm).

TABLE 7.3

Beating the Blues: The REML Estimates of the Treatment Effect Under Different Scenarios of MNAR Imputations

Scenario	Class	δ	Estimate	Stand Error	95% Confidence Interval	
					Lower	Upper
1	Constant	0,5	−2,62	1.72	−6,05	0,80
2		1	−2,23	1.74	−5,70	1,24
3		2	−1,45	1.80	−5,03	2,12
4	Increasing	0,5	−1,76	1.77	−5,65	1,28
5		1	−1,38	1.79	−5,95	2,20
6	Increasing then	0,5 (up to $k^* = 2$)	−2,32	1.74	−5,78	1,13
7	constant	1 (up to $k^* = 2$)	−1,64	1.78	−5,18	1,90

MI analyses used 50 imputations.

> 2. *Increasing then constant change*: Imputed values are increased up to time point k^* by $k \times \delta$, $k \le k^*$ and then remain constant at rate $k^* \times \delta$ for time points after k^*.

These scenarios represent a constant, linear increase, and linear increase then constant deviation from MAR, respectively. The results of the sensitivity analysis based on 50 imputations are presented in Table 7.3. The analysis Equation 7.1 without interactions is fitted to each imputed dataset after which the treatment effect estimates are pooled to form a single inference. It is apparent that even a small deviation from the MAR assumption (see scenario 1) makes the treatment effect insignificant at the 5% level. We therefore have to conclude that, after sensitivity analysis with the three plausible scenarios, there is hardly any evidence that the Beating the Blues treatment is superior to treatment as usual.

7.2 Assignments

7.2.1 Assignment

Repeat the analysis in this chapter, but leave the variance-covariance **V** matrix unstructured. Compare the results you found with the results found in this chapter. Which strategy do you prefer and why?

7.2.2 Assignment

Consider the Beating the Blues data (SPSS system file: AncovaBTB.sav) and perform the complete cases analysis as follows:

a. Select only complete cases (i.e. patients who adhere the study proto-col) and transpose them to a long-format dataset.

b. Start with Equation 7.1 and determine the best variance-covariance structure for complete cases (use the basic guidelines of Chapter 4).

c. Considering a compound symmetry covariance structure, inves-tigate the *Time* × *Treat* interaction in Equation 7.1 using complete cases. Is the interaction effect significant at the 5% level?

d. Exclude the *Time* × *Treat* interaction from Equation 7.1 and fit the simplified model to complete cases.

e. Compared the results of part d with Table 7.2 and discuss possible discrepancies.

7.2.3 Assignment

Consider the Beating the Blues data (SPSS system file: LongBTB.sav) and perform the following analyses:

a. Discuss the plausibility of the MAR assumption by considering the dependency of missingness at time *t* with *Duration*, *Depress* before t, *bdipre*, and the interactions *Depress* × *Treat*, and *Duration* × *Treat* at each time point ($t = 2, 3, 5, 8$ months).

b. Show by using a numerical example that the MNAR situation is still possible

c. Specify and motivate the multilevel regression model. Interpret the relevant regression parameters.

d. What variance-covariance structure of the residuals would you select before fitting the model (use the basic guidelines mentioned in Chapter 4)?

e. Discuss the results of the analysis.

f. Compare the ANCOVA and the change-score approach and discuss the discrepancy of the results.

References

Afshartous, David., & Wolf, Michael (2007). Avoiding 'data snooping' in multilevel and mixed effects models, *Journal of the Royal Statistical Society: Series A*, 170(4), 1035–1059.

Altman, D. (1985). Comparability of randomised groups. *Journal of the Royal Statistical Society. Series D (The Statistician)*, 34(1). https://doi.org/10.2307/2987510

Altman, D.G. (1991). *Practical Statistics for Medical Research*. London, UK: Chapman & Hall.

Anderson, Sharon Roe, Ariane, Auquier, Walter, W. Hauck, David, Oakes, Walter, Vandaele, Herbert, I. Weisberg (1980). *Statistical Methods for Comparative Studies: Techniques for Bias Reduction* (1st edn), New York: Wiley.

Andridge, R.R., & Little, R.J.A. (2010). A review of hot deck imputation for survey non-response. *International Statistical Review*, 78, 40–64.

Audigier, V., White, I.R., Jolani, S., Debray, T.P., Quartagno, M., Carpenter, J., van Buuren, S., & Resche-Rigon, M. (2018). Multiple imputation for multilevel data with continuous and binary variables. *Statistical Science*, 33(2), 160–183.

Bartlett, J.W., Seaman, S.R., White, I.R., & Carpenter, J.R. 2015. Multiple imputation of covariates by fully conditional specification: Accommodating the substantive model. *Statistical Methods in Medical Research*, 24(4): 462–487.

Beerens, Hanneke C., Zwakhalen, Sandra M.G., Verbeek, Hilde, Tan Frans, E.S., Jolani, Shahab, Downs, Murna, de Boer, Bram, Ruwaard, Dirk & Hamers, Jan P.H. (2018). The relation between mood, activity, and interaction in long-term dementia care. *Aging & Mental Health*, 22(1), 26–32.

Bethlehem, J.G. (2002). Weighting adjustments for ignorable nonresponse. In Groves, R.M., Dillman, D.A., Eltinge, J.L., & Little, R.J.A. (Eds.), *Survey Nonresponse* (chapter 18; pp. 275–287). New York: John Wiley & Sons.

Biering, K., Hjollund, N.H., & Frydenberg, M. (2015). Using multiple imputation to deal with missing data and attrition in longitudinal studies with repeated measures of patient-reported outcomes. *Clinical Epidemiology*, 7, 91.

Bland, M. (2000). *An Introduction to Medical Statistics* (3rd edn). Oxford, UK: Oxford University Press.

Boer de Michiel, R., Waterlander, Wilma E., Kuijper, L.D.J., Steenhuis, Ingrid H.M., & Twisk, Jos W.R. (2015). Testing for baseline differences in randomised controlled trials: An unhealthy research behaviour that is hard to eradicate. *International Journal of Behavioral Nutrition and Physical Activity*, 12, 4. https://doi.org/10.1186/s12966-015-0162-z

Brekelmans, M., Wubbels, T., & Créton, H.A. (1990). A study of student perceptions of physics teacher behaviour. *Journal of Research in Science Teaching*, 27, 335–350.

Carpenter, J.R., & Kenward, M.G. (2008). Missing Data in Clinical Trials—A Practical Guide. National Institute for Health Research, Publication RM03/JH17/MK: Birmingham, Available at: http://www.pcpoh.bham.ac.uk/publichealth/methodology/projects/RM03_JH17_ MK.shtml

Carpenter, J., & Kenward, M. (2013). *Multiple Imputation and Its Application*. New York: Wiley.

Casella, G., & Berger, R.L. (2002). *Statistical Inference* (2nd edn). Pacific Grove: Duxbury Press.

Collins, L.M., Schafer, J.L., & Kam, C.M. (2001). A comparison of inclusive and restrictive strategies in modern missing data procedures. *Psychological Methods*, 6, 330–351.

Cochran, William G. (1983). *Planning and Analysis of Observational Studies*, New York: Wiley.

Daamen, Marielle AMJ, Brunner-la Rocca, Hans Peter, Tan, Frans E.S., Hamers, Jan P.H., & Schols, Jos M.G.A. (2017). Clinical diagnosis of heart failure in nursing home residents based on history, physical exam, BNP and ECG: Is it reliable? *European Geriatric Medicine*, 1, 59–65.

Delattre, Maud, Lavielle, Marc, & Poursat, Marie-Anne. (2014). A note on BIC in mixed-effects models. *Electronic Journal of Statistics*, 8, 456–475.

Diggle, P., Heagerty, P., Liang, K.Y., & Zeger S. (2013). *Analysis of Longitudinal Data* (2nd edn). Oxford University Press.

Draper, N.R., & Smith, H. (1966). *Applied Regression Analysis*. New York: Wiley.

Eltinge, J.L., & Little, R.J.A., editors, *Survey Nonresponse* (chapter 18; pp. 275–287). New York: John Wiley & Sons.

Enders, C.K. (2010). *Applied Missing Data Analysis*. New York: Guilford Press.

Everitt, B.S. (1994). Exploring multivariate data graphically: A brief review with examples. *Journal of Applied Statistics*, 21(3), 63–94.

Field, A. (2018). *Discovering Statistics Using IBM SPSS Statistics*. Los Angeles: Sage.

Fitzmaurice, G.M., Laird, N.M., & Ware, J.H. (2012). *Applied Longitudinal Analysis*. New Jersey: John Wiley & Sons.

Gabriel, K.R. (1969). Simultaneous test procedures- Some theory of multiple comparisons. *Annals of Mathematical Statistics*; 40, 224–250.

Graham, J.W., Olchowski, A.E., & Gilreath, T.D. (2007). How many imputations are really needed? Some practical clarifications of multiple imputation theory. *Preventive Science*, 8(3), 206–213.

Greenberg, B.G. (1953). The use of analysis of covariance and balancing in analytical surveys. *American Journal of Public Health*, 43(6), 692–699.

Groenwold RH, White IR, Donders ART, Carpenter JR, Altman DG, Moons KG (2012). Missing covariate data in clinical research: When and when not to use the missing-indicator method for analysis. *Canadian Medical Association Journal*, 184(11): 1265–1269.

Gurka, M.J., Edwards, L.J., & Muller, K.E. (2011). Avoiding bias in mixed model inference for fixed effects. *Statistics in Medicine*, 30(22), 2696–2707.

Hsu, J.C. (1996). *Multiple Comparisons – Theory and Methods*. London: Chapman and Hall.

Hedeker, D., & Gibbons, R.D. (2006). *Longitudinal Data Analysis* (Vol. 451). New York: John Wiley & Sons.

Holland, P.W., & Rubin, D.B. (1983). *On Lord's paradox*. In H. Wainer & S. Messick (Eds.), *Principals of Modern Psychological Measurement* (pp. 3–25). Hillsdale, NJ: Erlbaum.

Hughes, R.A., White, I.R., Seaman, S.R., Carpenter, J.R., Tilling, K., & Sterne, J.A.C. (2014). Joint modelling rationale for chained equations. *BMC Medical Research Methodology*, 14, 28.

IBM Corp. Released 2020. *IBM SPSS Statistics, Version 27.0*. Armonk, NY: IBM Corp.

Ibrahim, J.G., & Molenberghs, G. (2009). Missing data methods in longitudinal studies: A review. *Test*, 18(1), 1–43.

James, G., Witten, D., Hastie, T., & Tibshirani, R. (2017). *An Introduction to Statistical Learning*. New York: Springer.

Johnson, R.A., & Wichern, D.W. (2014). *Applied Multivariate Statistical Analysis* (6th edn). London: Pearson.

Jolani, S. (2018). Hierarchical imputation of systematically and sporadically missing data: An approximate Bayesian approach using chained equations. *Biometrical Journal*, 60(2), 333–351.

Jolani, S., Debray, T.P., Koffijberg, H., van Buuren, S., & Moons, K.G. (2015). Imputation of systematically missing predictors in an individual participant data meta-analysis: A generalized approach using MICE. *Statistics in Medicine*, 34(11), 1841–1863.

Kayembe, M.T., Jolani, S., Tan, F.E., & van Breukelen, G.J. (2020). Imputation of missing covariate in randomized controlled trials with a continuous outcome: Scoping review and new results. *Pharmaceutical Statistics*, 19(6), 840–860.

Kayembe, M.T., Jolani, S., Tan, F.E., & van Breukelen, G.J. (2022a). Imputation of missing covariates in randomized controlled trials with continuous outcomes: Simple, unbiased and efficient methods. *Journal of Biopharmaceutical Statistics*, 32(5), 717–739.

Kayembe, Mutamba T., Jolani, Shahab, Tan, Frans E.S., van Breukelen, Gerard J.P. (2022b). Dealing with outcome and covariate missingness in randomized controlled trials: Comparison of simple with advanced methods. *submitted*.

Kenward, M.G., & Molenberghs, G. (2009). Last observation carried forward: A crystal ball. *Journal of Biopharmaceutical Statistics*, 19(5), 872–888.

Kleinbaum, David G., Kupper, Lawrence L., Nizam, Azhar. (2008). *Applied Regression Analysis and Other Multivariable Methods* (4th edn). London: Thomson higher education.

Konietschke, F., & Brunner, E. (2013). Are multiple contrast tests superior to the ANOVA. *The International Journal of Biostatistics*, 9(1): 63–73.

Landau, S., & Everitt, B.S. (2004). *A Handbook of Statistical Analysis Using SPSS*. New York: CRC Press.

Lane, P. (2008). Handling drop-out in longitudinal clinical trials: A comparison of the LOCF and MMRM approaches. *Pharmaceutical Statistics: The Journal of Applied Statistics in the Pharmaceutical Industry*, 7(2), 93–106.

Lewis, D K (1973) *Counterfactuals*. Cambridge: Harvard University Press.

Liu, Chunyan, Cripe, Timothy P., Kim, Mi-Ok. (2010). Statistical issues in longitudinal data analysis for treatment efficacy studies in the biomedical sciences, *The American Society of Gene & Cell Therapy*, 18(9), 1724–1730.

Liu, J., Gelman, A., Hill, J., Su, Y.-S., & Kropko, J. (2014). On the stationary distribution of iterative imputations. *Biometrika*, 101, 155–173.

Little, R.J. A. (1988). A test of missing completely at random for multivariate data with missing values. *Journal of the American Statistical Association*, 83, 1198–1202.

Little, R.J. (1994). A class of pattern-mixture models for normal incomplete data. *Biometrika*, 81(3), 471–483.

Little, R.J. (1995). Modeling the drop-out mechanism in repeated-measures studies. *Journal of the American Statistical Association*, 90(431), 1112–1121.

Little R. J., & Rubin D. B. (1987). *Statistical analysis with missing data*. New York: Wiley. First edition.

Little, R.J., & Rubin, D.B. (2020). *Statistical Analysis With Missing Data* (3rd edn). New York: Wiley.

Meinert, Curtis (2009). *Clinical Trials: Design, Conduct and Analysis*. New York: Oxford Scholarship Online.

Mohanraj, J., & Srinivasan, M.R. (2015). Assessment of covariance structure in longitudinal analysis. *Statistica and Applicazioni*, 13, 121–136.

Molenberghs, G., & Kenward, M. (2007). *Missing Data in Clinical Studies* (Vol. 61). New York: John Wiley & Sons.

Molenberghs, G., Fitzmaurice, G., Kenward, M.G., Tsiatis, A. and Verbeke, G. (Eds.) (2015). *Handbook of Missing Data Methodology*. London: CRC Press.

Moons, K.G., Donders, R.A., Stijnen, T., Harrell, F.E., Jr (2006). Using the outcome for imputation of missing predictor values was preferred. *Journal of Clinical Epidemiology*, 59, 1092–1101.

Murray, J.S. (2018). Multiple imputation: A review of practical and theoretical findings. *Statistical Science*, 33(2), 142–159.

Nieboer, A.P., Schulz, R., Matthews, K.A., Scheier, M.F., Ormel, J., & Lindenberg, S. (1998). Spousal caregivers' activity restriction and depression: A model for changes over time. *Social Science & Medicine*, 47, 1361–1371.

Ouwens, Mario J.N., Tan, Frans E.S., & Berger, Martijn P.F. (2001). Local influence to detect influential data structures for generalized linear mixed models. *Biometrics*, 57(4):1166–1172.

Potthoff, R.F., & Roy, S.N. (1964). A generalized multivariate analysis of variance model usefully especially for growth curve problems. *Biometrika*, 51(3), 313–326.

Proudfoot, J., Goldberg, D., Mann, A., Everitt, B., Marks, I., & Gray, J. (2003). Computerized, interactive, multimedia cognitive behavioural therapy reduces anxiety and depression in general practice: A randomized controlled trial. *Psychological Medicine*, 33, 217–227.

Raghunathan, T., Berglund, P.A., & Solenberger, P.W. (2018). *Multiple Imputation in Practice: With Examples Using IVEware*. Boca Raton: CRC Press.

Raghunathan, T.E., Lepkowski, J.M., Hoewyk, J.V., & Solenberger, P. (2001). A multivariate technique for multiply imputing missing values using a sequence of regression models. *Survey Methodology*, 27, 85–95.

Reilly, M. (1993). Data analysis using hot deck multiple imputation. *The Statistician*, 42, 307–313.

Resche-Rigon, M., & White, I. (2018). Multiple imputation by chained equations for systematically and sporadically missing multilevel data. *Statistical Methods in Medical Research*, 27(6), 1634–1649.

Rhoads, Christopher H. (2012). Problems with tests of the missingness mechanism in quantitative policy studies. *Statistics, Politics, and Policy*, 3(1), Article 6. https://doi.org/10.1515/2151-7509.1012

Robins, J.M., & Gill, R.D. (1997). Non-response models for the analysis of non-monotone ignorable missing data. *Statistics in Medicine*, 16(1), 39–56.

Rosenbaum, P.R. (2002). *Observational Studies* (2nd edn), New York: Springer.

Royston, P. (2004). Multiple imputation of missing values. *Stata Journal*, 4, 227–241.

Rubin, D.B. (1974). Estimating causal effects of treatments in randomized and nonrandomized studies. *Journal of Educational Psychology*, 66, 688–701.

Rubin, D.B. (1976). Inference and missing data. *Biometrika*, 63, 581–592.

Rubin, D.B. (1977). Assignment to treatment group on the basis of a covariate. *Journal of Educational Statistics*, 2, 1–26.

Rubin, D.B. (1987). *Multiple Imputation for Nonresponse in Surveys*. New York: John Wiley & Sons.

Rubin, D.B. (1996). Multiple imputation after 18+ years. *Journal of the American Statistical Association*, 91(434), 473–489.

Rubin, D.B. (2006). *Matched Sampling for Causal Effects*. Cambridge: Cambridge University Press.

Sauder, Derek C., & DeMars, Christine E. (2019). An updated recommendation for multiple comparisons. *Association for Psychological Science*, 2(1), 26–44.

SAS Institute Inc. (2016). *SAS® 9.4 Functions and CALL Routines: Reference* (5th edn). Cary, NC: SAS Institute Inc.

Schafer, J.L. (1997). *Analysis of Incomplete Multivariate Data*. London: Chapman and Hall.

Schafer, J.L. (2003). Multiple imputation in multivariate problems when the imputation and analysis models differ. *Statistica Neerlandica*, 57, 19–35.

Schulz, Kenneth F, Altman, Douglas G., & Moher, David (2010). CONSORT 2010 Statement: Updated guidelines for reporting parallel group randomised trials. *BMJ 2010*, 340. https://doi.org/10.1136/bmj.c332.

Seaman, S.R., Bartlett, J.W., & White, I.R. (2012). Multiple imputation of missing covariates with non-linear effects and interactions: An evaluation of statistical methods. *BMC Medical Research Methodology*, 12(1): 46.

Senn, S.J. (2006). Change from baseline and analysis of covariance revisited. *Statistics in Medicine*, 25, 4334–4344.

Siddiqui, O., Hung, H. J., & O'Neill, R. (2009). MMRM vs. LOCF: A comprehensive comparison based on simulation study and 25 NDA datasets. *Journal of Biopharmaceutical Statistics*, 19(2), 227–246.

Singer, J.D., & Willett, J.B. (2003). *Applied Longitudinal Data Analysis: Modeling Change and Event Occurrence*. New York: Oxford University Press.

Snijders, T.A.B., & Bosker, R.J. (2012). *Multilevel Analysis: An Introduction to Basic and Advanced Multilevel Modelling* (2nd edn). London: Sage Publisher.

Sullivan, T.R., White, I.R., Salter, A.B., Ryan, P., Lee, K.J. (2018). Should multiple imputation be the method of choice for handling missing data in randomized trials? *Statistical Methods in Medical Research*, 27, 2610–2626.

Speidel, M., Drechsler, J., & Jolani, S. (2020). The R Package hmi: A convenient tool for hierarchical multiple imputation and beyond. *Journal of Statistical Software*, 95(1), 1–48.

Steyerberg, E.W. (2009). *Clinical Prediction Models*. New York: Springer-Verlag.

Tan, F.E., Jolani, S., & Verbeek, H. (2018). Guidelines for multiple imputations in repeated measurements with time-dependent covariates: A case study. *Journal of Clinical Epidemiology*, 102, 107–114.

Tan, Frans E.S., Ouwens, Mario J.N., & Berger, Martijn P.F. (2001). Detection of influential observations in longitudinal mixed effects regression models. *Journal of the Royal Statistical Society: Series D*, 271–284.

Tan, F.E.S., Wuts, F., & Schols, I. (2008). Chapter 11: *Meervoudige Lineaire Regressie (Multiple Linear Regression)*, in Berger, M.P.F., Imbos, T.J., & Janssen, M.P.E., *Methodologie en statistiek 2*, Maastricht: Universitaire Pers Maastricht.

Tan, Frans E.S. (2013). Confounding in (non-) randomized comparison studies. *OA Epidemiology*, 1(3), 1–6.

van Breukelen, G (2006). ANCOVA versus change from baseline had more power in randomized studies and more bias in nonrandomized studies. *Journal of Clinical Epidemiology*, 59(9), 920–925.

van Breukelen, G (2013). ANCOVA versus change from baseline in non-randomized studies: The difference. *Multivariate Behavioural Research*, 48(6), 895–922.

van Buuren, S, Boshuizen, H.C., & Knook, D.L. (1999). Multiple imputation of missing blood pressure covariates in survival analysis. *Statistics in Medicine*, 18, 681–694.

van Buuren, S. (2018). *Flexible Imputation of Missing Data* (2nd edn). Boca Raton: CRC Press.

Verbeke, G., & Molenberghs, G. (2000). *Linear Mixed Model for Longitudinal Data*. New York: Springer-Verlag.

Verbeke, G., & Molenberghs, G. (2006). Short course: Longitudinal data, mixed models and incomplete data. Forest Research Institute, March 23–24. https// gbiomed.kuleuven.be/English/research/50000687/50000696/geertverbeke/ cursuskort/ldasc06nwjersey.pdf

Von Hippel, P.T. (2020). How many imputations do you need? A two- stage calculation using a quadratic rule. *Sociological Methods & Research*, 49(3), 699–718.

Vrieze, S.I. (2012), Model selection and psychological theory: A discussion of the differences between the Akaike Information Criterion (AIC) and the Bayesian Information Criterion (BIC), *Psychological Methods*, 17(2): 228–243.

Weakliem, David. (1999). A critique of the Bayesian information criterion for model selection. *Sociological Methods and Research*, 27: 359–397.

Weisberg, H.I. (1979). Statistical adjustments and uncontrolled studies. *Psychological Bulletin*, 86(5), 1149–1164.

White, I.R., & Carlin, J.B. (2010). Bias and efficiency of multiple imputation compared with complete-case analysis for missing covariate values. *Statistics in Medicine*, 29(28), 2920–2931.

White, I. R., Carpenter, J., Evans, S., & Schroter, S. (2007). Eliciting and using expert opinions about dropout bias in randomized controlled trials. *Clinical Trials*, 4(2), 125–139.

White, I.R., Carpenter, J., & Horton, N.J. (2012). Including all individuals is not enough: Lessons for intention-to-treat analysis. *Clinical Trials*, 9(4), 396–407.

White, I.R., Royston, P., & Wood, A.M. (2011). Multiple imputation using chained equations: Issues and guidance for practice. *Statistics in Medicine*, 30(4), 377–399.

White, I.R., & Thompson, S.G. (2005). Adjusting for partially missing baseline measurements in randomized trials. *Statistics in Medicine*, 24(7), 993–1007.

Widaman, Keith F (1985). Hierarchically nested covariance structure models for multitrait-multimethod data, *Applied Psychological Measurement*, 9(1), 1–26.

Wood, A.M., White, I.R., & Thompson, S.G. (2004). Are missing outcome data adequately handled? A review of published randomized controlled trials in major medical journals. *Clinical Trials*, 1(4), 368–376.

Zhu, J., & Raghunathan, T.E. (2015). Convergence properties of a sequential regression multiple imputation algorithm. *Journal of American Statistical Association*, 110, 1112–1124.

Index

Pages in *italics* refer figures and **bold** refer tables.

Printed in the United States
by Baker & Taylor Publisher Services